T0185286

Synthesis Lectures on Artificial Intelligence and Machine Learning

Series Editors

Ron Brachman, Jacobs Technion-Cornell Institute, New York, USA

Francesca Rossi, Thomas J Watson Research Center, IBM Research AI, Yorktown Heights, NY, USA

Peter Stone, University of Texas at Austin, Austin, TX, USA

This series publishes short books on research and development in artificial intelligence and machine learning for an audience of researchers, developers, and advanced students.

Vaishak Belle

Toward Robots That Reason: Logic, Probability & Causal Laws

 Springer

Vaishak Belle
School of Informatics
University of Edinburgh
Edinburgh, UK

ISSN 1939-4608   ISSN 1939-4616 (electronic)
Synthesis Lectures on Artificial Intelligence and Machine Learning
ISBN 978-3-031-21005-1   ISBN 978-3-031-21003-7 (eBook)
https://doi.org/10.1007/978-3-031-21003-7

© The Editor(s) (if applicable) and The Author(s), under exclusive license to Springer Nature Switzerland AG
2023
This work is subject to copyright. All rights are solely and exclusively licensed by the Publisher, whether the whole
or part of the material is concerned, specifically the rights of translation, reprinting, reuse of illustrations, recitation,
broadcasting, reproduction on microfilms or in any other physical way, and transmission or information storage
and retrieval, electronic adaptation, computer software, or by similar or dissimilar methodology now known or
hereafter developed.
The use of general descriptive names, registered names, trademarks, service marks, etc. in this publication does
not imply, even in the absence of a specific statement, that such names are exempt from the relevant protective
laws and regulations and therefore free for general use.
The publisher, the authors, and the editors are safe to assume that the advice and information in this book are
believed to be true and accurate at the date of publication. Neither the publisher nor the authors or the editors give
a warranty, expressed or implied, with respect to the material contained herein or for any errors or omissions that
may have been made. The publisher remains neutral with regard to jurisdictional claims in published maps and
institutional affiliations.

This Springer imprint is published by the registered company Springer Nature Switzerland AG
The registered company address is: Gewerbestrasse 11, 6330 Cham, Switzerland

Preface

Artificial Intelligence (AI) is widely acknowledged as a new kind of science that will bring about the next technological revolution. The vast majority of exciting reports that come our way about the use of AI in applications, however, are concerned with a very narrow technological capability: predicting future instances based on previously observed data. But AI, as understood by both scientists and science fiction writers, is clearly much broader.

This book is on the science of general-purpose, open-ended computational entities that deliberate and learn. Although such an agenda raises numerous philosophical and technical concerns that have no easy answers, some ideas have emerged that are attempting to tackle fundamental representational and reasoning problems for rational agents operating in complex, uncertain environments. The book builds on two such major developments in AI:

(a) the longstanding goal of integrating logic and probability for commonsense reasoning over noisy data; and
(b) theories of actions, dynamic laws and planning to achieve objectives in a changing world.

To that end, it presents the mathematical machinery for a logical language that integrates quantifiers, probabilities, actions, plans and programs.

Indeed, the unification of logic and probability has enjoyed a lot of attention in mathematics, logic, computer science and game theory. In AI, for example, areas such as statistical relational learning, neuro-symbolic systems, probabilistic databases, among others, are motivated by the need to incorporate noise and probabilistic uncertainty with logical knowledge and deductive machinery. Much of this work is limited to a *static* state of affairs, and so reasoning about a changing world via actions and plans is the next frontier. Many formalisms, moreover, limit the expressiveness for computational reasons, but this leaves open what a general account of first-order logic, probability and actions looks like. That is what this book seeks to address.

Who is this book for? Graduate students and researchers in computer science, artificial intelligence, philosophy, logic, robotics and statistics in the least would find the material useful. For instance:

- In computer science, epistemic logic has provided a formal model to capture distributed systems, multi-agent systems, privacy, cryptography and security. The integration of probability demonstrates how uncertainty could be further captured.
- In philosophy, the unification of logic and probability has been of interest from the point of epistemology and language. It will be useful to see a working proposal of such a unification for problems in AI.
- In statistics, a great amount of effort goes into capturing appropriate assumptions as well as exploiting tractable properties when modelling time and dynamics. The development of a language where no effort is needed in understanding how the dependencies between variables might change over actions is a fresh perspective on representational matters. Tractable properties are also shown to emerge as a special case when the appropriate limitations hold.
- In robotics, there is an increasing interest in autonomy and enabling commonsense reasoning along with real-time behavior. The book's focus on arbitrary actions and programmatic abstractions for partially specified behavior against a first-order knowledge base might be indicative of what is possible with very expressive languages. Clearly, that level of expressiveness does not entail real-time behavior, so a middle ground needs to be sought. But a roboticist can now revisit the expressiveness-tractability tradeoff by better articulating what is possible with representational richness.

A background in first-order logic is all that is needed to go through the contents of this book. It might even be possible to grasp the thrust of chapters without any knowledge of formal logic, but the reading of equations will require some understanding of logic. (We provide a brief introduction to first-order logic in Chap. 3).

For general audiences, Chaps. 1 and 2 motivate the scientific program. For researchers in statistical relational learning, these chapters also position the technical work in the book against popular relational languages in the machine learning community.

For readers better acquainted with modal logic, Chap. 10 might be a more comfortable read. For researchers in automated planning, Chap. 8 on belief-level regression might be especially interesting. For researchers in agent programming, Chap. 9 on programmatic abstractions for agent design under uncertainty might be worthwhile.

I hope this book motivates you to think about the many beautiful ways in which logic and probability interact.

Edinburgh, UK   Vaishak Belle
September 2022

Acknowledgments

Almost the entirety of the technical material in this book is a result of collaborations that began during my time at the University of Toronto. I owe immense intellectual debt to my co-authors and mentors: Hector Levesque, Sheila McIlraith and Gerhard Lakemeyer.

My thoughts on statistical relational learning and how it connects to the scientific program advocated in this book was significantly shaped during my time at KU Leuven. I am grateful to the energetic group of co-authors and mentors I had there. I am especially grateful to Luc De Raedt.

Many students and colleagues at the University of Edinburgh helped me sharpen my thoughts on logic and learning. I am grateful for their insights and feedback.

Over the years, many of my peers from the knowledge representation and artificial intelligence community have become close friends. I want to thank them for their encouragement and camaraderie.

The writing of this book, as well as the research carried out within would not be possible without generous funding from several institutions: DAAD (Germany), NSERC (Canada), FWO (Belgium), UKRI (UK) and the Royal Society (UK).

This book is dedicated to my wife, Sukanya, and my daughter, Bodhi, for their unending love.

Edinburgh, UK   Vaishak Belle
September 2022

Contents

About the Author

Vaishak Belle Ph.D., is a Chancellor's Fellow and Reader at the School of Informatics, University of Edinburgh (UK). He is an Alan Turing Institute Faculty Fellow, a Royal Society University Research Fellow, and a member of the Royal Society of Edinburgh's Young Academy of Scotland.

He has co-authored over 50 scientific articles on AI, and along with his co-authors, he has won several best paper awards. At the University of Edinburgh, he directs a research lab on artificial intelligence, specialising in the unification of logic and machine learning. He joined the University of Edinburgh in 2016, after postdoctoral fellowships at the University of Toronto (Canada) and KU Leuven (Belgium).

Introduction

1

> *They don't have intelligence. They have what I call*
> *"thintelligence". They see the immediate situation. They think*
> *narrowly and they call it being focused. They don't see the*
> *surround. They don't see the consequences.*
>
> —**Michael Crichton**, *Jurassic Park*

Artificial Intelligence (AI) is widely acknowledged as a new kind of science that will bring about (and is already enabling) the next technological revolution. Virtually every week, exciting reports come our way about the use of AI for drug discovery, game playing, stock trading and law enforcement. And virtually all of these are mostly concerned with a very narrow technological capability, that of predicting future instances based on past instances. Identifying statistical patterns, correlations, and associations are, without doubt, extremely useful. In the first instance, they are needed in applications to inspect features and properties of interest in observed data. But AI, as understood by both scientists and science fiction writers, is clearly much broader. In fact, pattern recognition, machine learning and finding associations by mining data are closely related subfields of AI. Put differently, from first principles, what distinguishes big-data analysis from AI is that the set of capabilities we wish to enable with the latter. We are not interested in a "thintelligence", but rather a general-purpose, autonomous computational entity that, in the very least, has agency.

1.1   A Science of Agency, Deliberation and Learning

To develop a science of agency with deliberation and learning, we need to address several critical philosophical concerns, including:

- Which capabilities are of interest?
- What sort of framework allows us to capture those capabilities?

© The Author(s), under exclusive license to Springer Nature Switzerland AG 2023
V. Belle, *Toward Robots That Reason: Logic, Probability & Causal Laws*,
Synthesis Lectures on Artificial Intelligence and Machine Learning,
https://doi.org/10.1007/978-3-031-21003-7_1

- How are we to reason about the system's uncertainty about the world, and the laws that govern it?
- Which of those capabilities and laws can be codified, using mathematical language, and how is that language defined?

It should not come as a surprise that there is no consensus yet on how such questions should be answered. To wit, consider the simple capability of reasoning about and manipulating ordinary things, as might be expected of a robotic caretaker servicing an office, for example. It might be tasked with cleaning up rooms, delivering coffee to individuals issuing such requests, and so on. For one thing, uncertainty could range from disjunctive (e.g., either-or) to existential (e.g., there is someone with a certain property) to probabilistic (e.g., one event is more likely than the other), in addition to other notions. For another, if the functionality is to be addressed in a general way, a wide range of technical concerns arise. In the very least, consider:

- Should the system's behavior be learnt entirely from data, or only partially?
- If the latter, what knowledge does a system need to have in advance (e.g., provided by a modeler) versus what can be acquired by observations?
- What kind of semantics governs the updating of a priori knowledge given new and possibly conflicting observations?
- How does the system generalize from low-level observations to high-level structured knowledge?

Technical solutions to such concerns need to be further embedded in a society, where compliance with cultural and social norms is surely demanded. To reiterate our point above more bluntly, we are yet to identify any single framework or language that is shown to be appropriate for AI systems, understood so broadly.

Be that as it may, some ideas have emerged that are attempting to tackle fundamental representational and reasoning problems for rational agents operating in complex, uncertain environments. This book builds on two such major developments in AI:

- The longstanding goal of integrating logic and probability for commonsense reasoning over noisy data.
- And, models of actions and planning to achieve objectives in a changing world.

We believe that what is needed as a key driver towards general-purpose AI, such as autonomous open-ended robots, is a framework that unifies:

$$(logic \; + \; probability) \; + \; actions.$$

However, as is clear from our many open-ended questions above, a precise understanding of what level of expressiveness is needed for commonsensical rational agents is lacking.

This means we should strive for an application-independent, general-purpose language. A language that combines abstract, logical reasoning with probabilistic data. A language for reasoning about objects and their properties. A language that allows for reasoning about past events and hypothetical futures. A language that can express recursive plans of action. A language for refining knowledge with new but imprecise and noisy observations.

We motivate a proposal for such a unification in this book. We will begin by positioning historical developments, and then turn to the issues of big data and knowledge acquisition, which are important but orthogonal concerns for this book.

1.2   **Logic Meets Probability**

In the early days of AI, John McCarthy put forward a profound idea to realize artificial intelligence (AI) systems: he posited that what the system needs to know could be represented in a *formal language*, and a general-purpose algorithm would then conclude the necessary actions needed to solve the problem at hand. The main advantage is that the representation can be scrutinized and *understood by external observers*, and the system's behavior could be *improved* by making statements to it. Numerous such languages emerged in the years to follow, but first-order logic remained at the forefront as a general and powerful option. Propositional and first-order logic continue to serve as the underlying language for several areas in AI, including constraint satisfaction, automated planning, database theory, ontology specification, verification, and knowledge representation.

One of the main arguments against a logical approach is that in practice, there is pervasive uncertainty in almost every domain of interest: these can be in the form of measurement errors (e.g. readings from a thermometer), the absence of categorical assertions (e.g. smoking may be a factor for cancer, but cancer is not an absolute consequence for smokers), and the presence of numerous "latent" factors, including causes that the modeler may simply not have taken into account, all of which question the legitimacy of the model. The upshot is that on the one hand, logic was seen as an inappropriate tool, as it is "rigid" (sentences always evaluate to true or false), "brittle" (sentences in the knowledge base must be true in all possible worlds) and discrete (as opposed to the continuous error profiles for thermometers). On the other, the knowledge of the system, as posited in the declarative approach, may not only be incomplete but may be impossible to specify a priori (e.g., consider the many dimensions to telling a system on what constitutes as a face in high-resolution photographs).

Modeling uncertain worlds needed a rigorous formulation, and this came in the form of probabilistic models, including ones admitting a graphical representation, such as Bayesian networks. Such models allow one to effectively factorize the joint distribution over random variables. What makes such models particularly attractive is that both the probabilities of the variables in a given model, as well as the dependencies themselves can be learnt from data, thereby circumventing the requirement that the model needs to be provided by some omniscient modeler. Probabilistic models, obtained either by explicit specification or implicitly induced by means of modern machine learning methods such as deep learning, have supercharged the application of statistical methods in language understanding, vision and data analysis more generally.

Despite the success of probabilistic models, we observe that they are essentially propositional, but are nonetheless deployed in an inherently relational world. That is, they easily make sense of "flat" data, where atomic events are treated as independent random variables. But the environment that a robot operates has things in it: some objects may be inside another, others on top, some fragile and some heavy. We would need to reason about the properties of these things to manipulate and transport them successfully. Likewise, in medical records, it makes little sense to treat individual entries on patient symptoms as independent, since it ignores relationships between co-occurring symptoms, and family history. This encouraged the design of probabilistic concept languages, culminating in the area of statistical relational learning, neuro-symbolic AI and many other hybrid formalisms integrating probabilistic observations and high-level reasoning and/or planning. These formalisms borrow syntactic devices from finite-domain first-order logic to define complex interactions between random variables in large-scale models over classes and hierarchies.

With so many formalisms to choose from, which language shall we work with? The main thing to note is that, in often distinct ways, these languages are carefully designed to balance expressiveness versus computational efficiency for the application context at hand. However, as a result, we are left with limited languages that offer some benefits over propositional approaches but are overly restrictive for other concerns. A central problem with such probabilistic concept languages is their very controlled engagement with first-order logic: by almost exclusively considering finite-domain relational logic, succinct modelling may be admitted, but it is ultimately no more powerful than propositional logic from an expressiveness viewpoint. In some programmatic approaches, moreover, logical connectives such as disjunctions are also disallowed.

Interestingly, such probabilistic concept languages are drawn from earlier, more general, studies on unifying first-order logic and probability, such as the works of Nilsson, Bacchus and Halpern. McCarthy and Hayes, in fact, were the first to suggest the following:

(i) *It is not clear how to attach probabilities to statements containing quantifiers in a way that corresponds to the amount of conviction people have.*

(ii) *The information necessary to assign numerical probabilities is not ordinarily available. Therefore, a formalism that required numerical probabilities would be epistemologically inadequate.*

Translating such sentiments to a desiderata of sorts, one might say a general-purpose language should support full first-order logic, but also allow probabilistic assertions. In other words, it should allow a purely probabilistic specification, if the application demands it and the information available allows it. Analogously, such a language should allow a purely logical specification, if no probabilistic information is available. And, of course, everything in between: for example, it should be possible to have an initial database consisting of only first-order formulas, then gradually add purely probabilistic formulas, and obtain appropriate conclusions from that resulting database. Moreover, from such a general language, one may then determine which fragment is sufficient for the application at hand, and constrain the language accordingly. This is an important advantage with rich languages.

Naturally, the downside of working with such a powerful language is that we will not be able to say very much about efficient computation is every instance. With a specific fragment in mind, that is possible. But not in general. Since the game here is to really understand the principles and theory behind integrating logic and probability, we will accept the matter, and consider concrete computational strategies at a later stage.

Not surprisingly, we will be in a very similar position with regards to reasoning about world dynamics. We will aim for a general language, on the one hand, and focus on computation only with appropriate fragments.

1.3   Actions

Reasoning about events, actions, plans and programs has a long history in computer science and AI. Similar to the many proposals in the literature for commonsense reasoning, we have plenty of formalisms to choose from for capturing actions. Formalisms such as temporal logics allow us to reason about time, including the positioning of properties in the current and future states (e.g., the variable will never go above the value of 100). Markov process allows the stochastic modelling of sequential events. When coupled with a reward function, they can be used to compute the sequence of actions to be taken by an agent to maximize the overall reward. Planning languages such as STRIPS describe the current state of the system as a database, and by means of a synthesis algorithm, a sequence of actions can be produced that changes the state to a desired database.

Similar to the observation we just made about probabilistic logical representation languages, there are very many models of actions, with some limitation on what kinds of things can be expressed. As scientists consider more challenging applications, a new feature would be considered desirable to add, and inevitably a new modelling language would be introduced, with a corresponding semantics. Of course, there is no way we can completely future-proof a language against all possible desirable features. But at least we can consider a language that is powerful enough to reason about features such as:

- Causal laws relating actions and effects.
- Internal actions that can be performed by the robot to the change the world state, sensing actions that do not change the world state but only what the robot knows, and exogenous actions that affect the world but are performed without the robot's control (and possibly without its knowledge).
- Reasoning about the past, hypothetical and counterfactual events.

- Reasoning about the beliefs, desires and intentions of all the agents in the environment.
- Reasoning about discrete and continuous, noise-free and noisy actions and sensors.
- Expressing atomic actions, sequential plans, recursive plans, program-like plans, and partial instructions for the robot to execute.

What are after, then, is a unifying "theory of dynamics", and again, we turn to first-order logic but now extended for actions. In other words, it would be ideal if first-order logic provided the substrate that allows us to reason about both probabilities and actions, which would then count as a general proposal in line with our aims. As we shall see, such a possibility does exist, and is a fairly simple extension to the one of most popular knowledge representation languages: the situation calculus. Originally postulated by McCarthy, and later revised by Reiter, it has enjoyed considerable attention as an important knowledge representation language with extensions for time, plans, programs, inductive definitions, abstraction, rewards and high-level control. Basically, initial knowledge is a standard (unrestricted) first-order theory, over which we define actions and effects. Actions result in some formulas in the theory changing values, depending on which predicates are affected by an action. The key feature, like temporal and dynamic logics, however, and unlike dynamic Bayesian networks and planning formalisms, is that the underlying language allows us to reason about arbitrary trajectories of actions. So, one can reason about the past and the future.

So, the situation calculus has all the expressiveness of standard first-order logic together with a theory of actions. All of that is studied comprehensively in the introductory book by Raymond Reiter. What this book is about is further extending that framework for reasoning about probabilities in a general way.

1.4   Some Related Areas

Before going further, it might be useful to position some existing, well-established areas in the context of our discussion on first-order probabilistic languages. They help motivate the kind of generality we are aiming for.

- **Classical databases.** Databases are defined over a relational logical language and a finite set of constants, the latter denoting the individuals and values in the database. A database is equivalent to a finite and consistent set of atoms, with the understanding that all atoms not mentioned in that set are false.

  Transactional databases allow for the execution of *commands,* which amount of adding new atoms to the set, and deleting others.

  For example, a university database might consist of all of the enrolled students in the current year, along with their phone numbers and details of the courses they are undertaking. Matriculating a new student would mean the addition of this student to the set of enrolled students, and adding the relevant personal and course details. When a student graduates, they would be removed from the set of enrolled students.

- **Incomplete databases.** Uncertainty about the truth of atoms might mean we entertain disjunctive knowledge. For example, while scanning a handwritten text, we might be unsure about the first name of a student with ID #243: it could either be *Mary* or *May*. Each alternative together with the remaining facts correspond to a finite and consistent set of atoms, so the representation is allowing for multiple possible worlds.

- **Probabilistic databases.** Perhaps the student with ID #243 had to submit multiple forms, and although the handwritten text is problematic to scan in all of them, it might seem more probable that it is *Mary* rather than *May*. Probabilistic databases allow probabilities on atoms, and more generally on possible worlds.

- **Probabilistic relational languages.** To accord probabilities to both atoms and formulas, formalisms such as Markov logic networks and relational Bayesian networks have emerged. These can be seen to extend standard probabilistic formalisms such as Markov and Bayesian networks with a relational syntax allowing for an easier way to define intricate probabilistic models involving entities and their relationships. Alternatively, probabilistic logic programming allow modelers to decorate Horn rules with probabilities, to similar effect.

  For example, when examining electronic health records, it is useful to learn predictive models that can natively understand the relationships between patients, diagnoses, prescribed medications, family history and the progression of symptoms over months and years. Here, a probabilistic relational model can be built to leverage such relationships. Actions can be further defined for such models: for example, given probabilistic variables for the positions of objects, noisy actions that move these objects to other locations would affect the distributions of those variables. The distribution would need to account for the error profile of the actions to capture the probabilistic nature of the current position.

- **Automated planning.** Perhaps the simplest model for automated planning can be defined using classical databases. Such a database could denote the initial state, and actions such as moving from one location to another, picking up and dropping objects, and so on, can be seen as transactions over such databases amounting to the addition and deletion of facts.

  A more involved planning language might allow for uncertainty over the initial state (using multiple possible worlds), as well as involved actions that are *context-dependent*. For example, if the floor is slippery, a move action may cause the robot to fall rather than just move ahead. Probabilities can be further accorded to the possible worlds as well as the unintended effects of actions, thereby necessitating the need for dynamic probabilistic relational languages.

  Usually a solution to an automated planning problem is simply a sequence of prescribed actions that transform the initial state to a desired goal state. However, when there is uncertainty about the initial state, or about the type of observations that the agent might encounter, solutions can be iterative or even recursive, resembling a program with if-then-else and while loops.

Although there are numerous formalisms in each of these areas, as can be surmised from the discussion the underlying mathematical language is generally built from fragments of logic and/or probability and/or actions. In that regard, what we are aiming for in this book is a single logical language that allows for probability theory and all of first-order logic embedded in a rich theory of actions. Thus, it is very general.

1.5   Computation, Big Data, Acquisition and Causality

There a few key dimensions not addressed in this book.

Computation: Let us start with the most significant, that of efficient reasoning. Given the observation that there are plenty of languages making the expressiveness-computation tradeoff in the literature, why opt for a first-order language, which we know to be semi-decidable? (In fact, we will even have some second-order features, which makes matters more challenging.)

As already discussed, most of the popular ones are propositional, and even this does not make them "tractable", because propositional reasoning is NP-complete, and reasoning about possible propositional worlds is #P-complete. On the one hand, with a first-order language, simple things like applying a finite-domain assumption yield a propositional language, so such a move is always possible. On the other hand, there are plenty of other strategies for making first-order fragments tractable (e.g., Horn logic), and some of these may prove more attractive from a representation viewpoint than propositional logic. Reasoning matters aside, then, the issue is really one about representation. First-order and second-order logic are widely regarded as the formal language of choice for representing quantified knowledge, and it does not seem prudent to abandon this so early in the game of designing general-purpose cognitive architectures. Moreover, none of the statistical relational languages really support reasoning about past and future events, and none of the temporal and dynamic logics really support first-order probabilistic reasoning. So there is a very urgent need for designing a language like the one in this book.

With that in mind, what are we able to say about computation? Many things, as it turns out. For example, in later chapters, we consider two popular strategies for reasoning about actions in the situation calculus literature called regression and progression. Regression reduces a query against a sequence of actions to a query about initial knowledge. Progression updates the entire initial knowledge against a sequence of specified actions, much like a database update. We show both of these can be extended to probabilistic settings, which is another way of saying that we only need to care about a reasoning module for a first-order probabilistic theory without actions. A candidate from statistical relational learning could very well be considered here, for example, as could a Bayesian network. Moreover, we also show that the general theory admits the implementation of a programming language, defined over a convenient fragment, where Bayesian inference is all you need. Other fragments could be considered instead, and this is precisely the benefit of exploring results in a general way.

Big data and acquisition: It seems unthinkable nowadays to write a book about AI that does not mention the word "data" (or even worse, "big data") on every page. Pattern recognition, machine learning and data mining are essentially about identifying statistical associations in observed data. It is clear that any intelligent agent will need this capability. In fact, just as deduction goes back to Aristotle, who suggested suggested the use of *syllogisms* as a means to arrive at new and necessary truths, so too does *induction*. Aristotle suggested, for example that we need the means for generalizing an argument *from the particular to the universal*. That is, deduction produces new knowledge in the form of conclusions not explicitly stored in the knowledge base as beliefs, and induction produces new knowledge via statements evidenced only in individual instances. But this knowledge is not necessary in the sense of being irrefutable. Rather than seeking a generalization that explains all the observations, we might construct a sentence such that the support for this sentence is more than the support for other possible generalizations. The support might be a probabilistic function over the observations, for example.

Given this, we clearly need to investigate strategies for inducing knowledge bases, in the very least. In fact, most statistical relational learning formalisms are geared for learning relational associations. Inducing the grandparent predicate from observations about atoms of parents and parents of parents, for example. Integrating such capabilities with the formal language of this book is, therefore, an important yet orthogonal direction to explore. (Note that since learning limited languages is easier than more expressive ones, the matter is not fully divorced from our choice of the representation language. But if the representation language is too weak, we simply cannot get the robot to reason about the world the way we wish for it too, and so we have to pay the price somewhere.)

It is worth briefly noting that there are non-trivial issues related to the representation language in the context of acquisition from data. They are largely related to the so-called *symbol grounding problem*. For example, how do we link predicates in the representation language to concrete entities in the real world? After all, symbols must be understood in terms of something more than just other symbols. For example, the predicate *Wall* might refer to an actual wall in the vicinity of a robot; by extension, objects on that wall, such as a painting or a window, need corresponding representations in the language. Analogously, how can we extract and reason with predicates of the appropriate granularity given observational data? A concrete entity (such as a dog) might be understood in terms of its atomic composition, its shape, its relation to other agents, and so on. Ultimately, such questions are asking a fundamental question: *where do symbols come from?*

There is a lot to be said about such concerns, and addressing the symbol grounding problem from a conceptual or a pragmatic angle continues to be an active research area, although many workable proposals exist. For the purposes of this book, it is also seen as an orthogonal issue.

Causality: However, we should be careful to not assume that there is a simple way to learn a full theory purely from observed data. For example, the data might suggest that every time the mug is broken, the robot has also executed a throw action, but not that the throwing is causing the mug to break. Likewise, data might suggest that individuals who smoke are reported to also have asthma, but it may not say that smoking is a factor for asthma, or that reducing or even stopping smoking might alleviate the condition.

Knowledge about the world is more than just associations, and is often causal. A heavy object placed on a glass table causes the table is break; fire can be sustained by adding paper or wood, but not if there is no oxygen; on earth, dropping objects from the top of a building causes it to fall to the ground; two trains starting at the same time and traveling at the same speed from London to Manchester will get to the destination at the same time, unless they stop along the way at stations for different lengths of time; and so on.

In sum, knowledge about cause-effect is essential to understand how the world works. It is with this type of knowledge that one can answer why (e.g., "why did the fire sustain for so long in the office?") and what-if (e.g., "if we had installed a fire resistant door for the kitchen, would the fire still have spread to the rest of the office?") questions. Conventional machine learning does not focus on such questions, but there are, fortunately, strategies for inducing action models and causal dependencies. Clearly, we would need to couple such strategies together with association rule learning to obtain a theory of the sort we consider in this book. But we argue that these issues are orthogonal, and the focus in the first instance must be on the language to model the agent's capabilities. In a statement remarkably similar in spirit, Judea Pearl writes:

This is why you will find me emphasizing and reemphasizing notation, language, vocabulary and grammar. For example, I obsess over whether we can express a certain claim in a given language and whether one claim follows from others. My emphasis on language also comes from a deep conviction that language shapes our thoughts. You cannot answer a question that you cannot ask, and cannot ask a question that you have no words for.

And, as with Pearl and the knowledge representation community more generally, we will identify with "representation first, acquisition second."

One final remark could be made on the topic of cause-effect and causality. In this book, as is standard with action and planning formalisms, we are focusing on physical actions such as moving a robot, enrolling a student and dropping an object. Causality, as often referred to the scientific community, is typically centered around examples involving medical interventions, such as the effect of administering a drug, or others from econometrics. These examples are of the type-level or *general causality*, and can be contrasted to token-level or *actual causality*, where we are interested in the actions that led to an event. It should not come as a surprise there is a close relationship between actual causality and the type of theories we develop in this book, and an influential causal framework, called *structural equation models*, has parallels to theories of action. We refer interested readers to the bibliographic section of this chapter for more information, and do not elaborate on it further in this book. We will only note that reasoning about causality is an important ingredient for a commonsensical rational agent, and here too, a first-order theory of action is attractive to be able to talk about objects, their properties and narratives based on trajectories.

1.6   Key Takeaways

Reflecting on the main arguments we have made above, let us conclude this chapter with the key takeaways on what to expect with this book:

- The book presents a formal language to unify logical reasoning, probabilistic reasoning and reasoning about actions. Although there has been considerable attention on these issues, this book is not about the philosophical aspects underlying such a unification, but rather on a general and workable mathematical model. In particular, we provide tools and ideas on how to make such a language useful for practical problems in AI, from reasoning about hypothetical queries to programming.
- This book follows the motto of "representation first, acquisition second", and focuses only the representation. This is justified, as we shall see, given the nuances in the interaction between first-order logic, probabilistic assertions, noise and actions.
- It is built on first-order logic (more precisely, its a dialect with some second-order logical features), and so inherits the properties and expressive power of first-order logic. Classical first-order and second-order logic is widely and extensively studied, and serves as the foundation for many formal inquires on computation and mathematics.

  In particular, the situation calculus, which is what the book is based on, is a first-order language that allows predicates and functions to express properties that are true, not only for the current history of actions, but all possible histories. This makes the language less like planning languages but more like branching-time logics. On the one hand, the language provides an avenue to discussing thorny issues such as the frame, qualification and ramification problems, which are awkward to discuss with state-based abstractions (such as transition systems). On the other, as has already been established, the language allows noteworthy and natural extensions for concepts such as discrete and continuous time, decision theory, goal change, agent ability, belief revision, concurrency, interleaving deliberation and observations, actual causality, probability (this book) and so on.

  At any such history, then, the world is described using first-order formulas, which makes it significantly more expressive that databases, for example, since we do not restrict the domain. This allows "open-world" modeling.
- Probabilistic logics define belief in terms of measures on possible worlds, and in our setting, we will define measures on first-order tree-like worlds. That is, the worlds interpret first-order formulas for every possible history of actions.

  Even leaving aside actions, the language for probabilistic knowledge, then, is significantly more expressive than probabilistic propositional languages and finite-domain probabilistic relational languages commonly seen in the literature.

- The integration of probabilities with the language does not force the model to commit to a single distribution, and is entirely determined by the constraints in the knowledge base. This means that one can capture fully probabilistic knowledge (e.g., a Bayesian network), fully logical knowledge (e.g., a first-order theory), and everything in between (e.g., a first-order theory with noisy observations), making it a powerful language for modeling incomplete information.
- Probabilistic belief, classically, is an extension of categorical belief (e.g., "I know this, and not that"), but conventional languages often do not engage with such modalities explicitly, and simply provide a means to reason about conditional probabilities. Our language will include an explicit epistemic operator, which is useful to reason about meta-beliefs (e.g., "I know that I do not know the temperature, outside so I should find out whether I need a sweater by looking through my window") and the beliefs of other agents (e.g., "I know that Bob does not know my telephone number").

Be that as it may, it is worth emphasizing that although a number of representational matters are addressed in the book, we are only scratching the surface when tackling general-purpose agents. As is clear from the overall motivation, the representational issues studied here are far from exhaustive. Modelling time, resources, rewards, norms, and so on are important topics, and how they interoperate with a probabilistic language needs to be considered carefully. Moreover, when it comes to the engineering challenges in building agents that function in the real world, everything from tractability to learning to symbol grounding to response time to the cognitive architecture are fundamental meta-level concerns. A book such as ours focusing on the representation language does not go into how such challenges and possible design choices are to be addressed. It simply clarifies the interaction between knowledge, probability, action and perception, nothing more. The physical instantiation of the modelling language and its use in real-time control is left open.

As argued earlier, the focus and study of the representation language is by choice, to identify and isolate issues with knowledge and observations in an expressive setting. Its application for building, say, an actual robot is out of scope for this book. Nonetheless, we provide ample references and discussions on where one may find less expressive variants deployed in practice.

As a final remark, some readers of this book may not necessarily wish to work with the situation calculus. And indeed, the language of the situation calculus is not a critical factor for integrating logic, probability and action. There are plenty other proposals on the subject. So what can the reader gain from this book? The following things are clear: whatever the language of choice might be for the reader, it needs to involve reasoning about possible worlds. To reason about all possible histories, these worlds need to be tree-like structures, and to allow for first-order expressiveness, the nodes in the tree need to be able to interpret relations and functions. We need to provide mechanisms for reasoning about future queries, and be able to articulate plans and programs. At the time of writing, such a gamut of results have not been established for any other language, at least one at the level of generality and expressiveness afforded by the situation calculus. If the reader were to consider other sorts of languages, obtaining such results might follow analogous formulations, and even if that were not the case, its worth understanding how such results work intuitively. A modal reconstruction of the situation calculus in the penultimate section yields a language that is perhaps easier to follow for those familiar with dynamic and program logics, and could be a starting point for the reader to define their own formal apparatus for logic + probability + actions.

1.7 Notes

John McCarthy's early proposal on designing AI systems using formal logic can be found in [166], which also suggested the first version of the situation calculus. Later, McCarthy and Hayes [167] identified philosophical and technical issues that might arise from the use of such a language, and articulated the problem with ascribing probabilities to logical formulas.

The variant of the situation calculus used in this book is the revised version from Reiter. Reiter's book provides a comprehensive treatment on the foundations of the situation calculus, as well as some major extensions to it [190]. While time, plans and programs are discussed in that book, other concepts such as inductive definitions, abstraction, program synthesis, verification and causality have also been looked into for the situation calculus in the recent years; see, for example, [10, 13, 47, 138, 147, 149, 213]. In particular, the probabilistic variant from the current book is based on the work by Bacchus, Halpern and Levesque (BHL henceforth) [8]. Its extension for continuous probabilities, among other things, that make up the other chapters in the book can be found in [22–28].

An introduction to probabilistic graphical models such as Bayesian networks can be found in [178]. The use of probabilistic methods in machine learning is explored in [173]. For an introductory book on the use of probabilistic methods in robotics, see [219].

For early work on unifying logic and probability, see [5, 82, 100, 174]. Less expressive variants have been explored within the communities of statistical relational learning [70, 91, 187, 192, 193], probabilistic databases [210], neuro-symbolic AI [15, 83, 201] and high-level control [144, 155, 169, 183]. In virtually all of these communities, there is an urgent need for integrating low-level perception with high-level expert knowledge.

The intractability of (exact) probabilistic inference, especially in the context of the propositional languages considered in statistical relational learning, is discussed in [6]. Approaches to make first-order reasoning tractable include proposals such as [17, 135].

For discussions and development on the symbol grounding problem, see [93, 108, 209, 212] and references therein. For a discussion on automated planning and various types of solutions, see [85]. For works on dynamic probabilistic relational models, see, for example, [175, 221]. For algorithms on iterative and recursive plans, see [116, 149, 208].

Judea Pearl's statement about "representation first, acquisition second" is taken from [180], which also serves as a gentle introduction to concepts in causality. For more technical introductions, including on actual causality, see [102, 113, 179]. For the connection between structural equation models and theories of action, see [13, 77, 114, 115, 123, 230]. In particular, [13, 115] study the benefits of using first-order logic for causality. For intuitive arguments that reasoning about causality is important for general-purpose AI, see [163, 180].

Representation Matters

2

> *"But it is a pipe." "No, it's not," I said. It's a drawing of a pipe. Get it? All representations of a thing are inherently abstract. It's very clever.*
>
> **—John Green, The Fault in Our Stars**

In this chapter, we will informally reflect on the interplay between first-order logic and probabilities. We will then introduce the language formally in the subsequent chapters.

2.1   Introduction

Our interest in this book is on the representation of information—capturing the knowledge of a robot, say—and how that representation churns out interesting conclusions that determine what the robot knows and what it should do next. As discussed in the previous chapter, we are glossing over the problem of *acquisition,* however important that may be (and it undoubtedly is), and focusing foremost of the *language* for representing information. Ensuring that this language is equipped with the tools for modelling and reasoning about the world is of fundamental import.

Which language then should we choose, and what should we be representing? Turing machines are abstract and foundational structures for investigating the computable universe. With only read and write operations, it is minimal and powerful, but precisely because it is a computational abstraction means that we have no means to concretely talk about what the robot knows about the environment that it is operating in. Likewise, binary system code is ultimately how computers will run robotic software. However, system code is clearly too coarse and opaque to explain a purposeful agent. Thus, we need to introduce a modelling framework that, on the one hand, is understandable to a domain expert and enables the updating and provision of new information. On the other hand, it lends itself to a computational process; in other words, we might contrast this requirement from vague, high-level

© The Author(s), under exclusive license to Springer Nature Switzerland AG 2023
V. Belle, *Toward Robots That Reason: Logic, Probability & Causal Laws*,
Synthesis Lectures on Artificial Intelligence and Machine Learning,
https://doi.org/10.1007/978-3-031-21003-7_2

instructions in natural language such as "get the robot to deliver coffee and mail across the floors of this building." The latter instruction is, clearly, well-intentioned, but provides no obvious strategy for building a robot.

By a *representation* we mean a mapping from one entity to another, the former being concrete and computationally processable. For example, *rain* might stand for the assertion that it is raining today. *Reasoning* is the computational processing of representations to produce new knowledge. In particular, in this book we focus on formal (symbolic) logic, where all information is stored explicitly or implicitly using symbols. The distinction between *explicit* and *implicit* is important: explicit knowledge is often the information directly modelled or provided by the domain expert—these could come in the form of rules, databases, knowledge bases, graphs, or any other structured data—and implicit knowledge is what is obtained from the explicit knowledge through one or more reasoning steps.

Reasoning, moreover, could be understood in two ways. In the first instance, it can define what the mathematical framework looks like for obtaining implicit knowledge from explicit knowledge. In logic, we often provide a semantics and/or a proof theory that says when sentences are *entailed* from a knowledge base (a set of sentences). In addition, reasoning could also provide an implementation strategy for computing entailments: an algorithm that takes as input a knowledge base and a query and outputs whether the query was entailed or not. Theorem provers, model checker, and satisfiability solvers, for example, are standard platforms for designing such algorithms. To a large extent, this book will mostly focus on the first facet of reasoning, but we will show that by applying certain restrictions, it will be possible to leverage well-known algorithms.

Returning now to the goal of the book, which is to unify logic, probability and action, let us first start by asking the following question. What is the desiderata for the unification? What does the representation language look like, and what can we say about how to reason with knowledge expressed in that language? In particular, given our interest in developing a general proposal, how would such a proposal compare to a the wide range of formalisms in the literature, such as those considered in statistical relational learning?

We will approach these questions from different perspectives, and hopefully convince the reader that we will be developing one of the most general formalisms to date. To do this, let us begin the discussion with an informal refresher on logic and probability, and then turn to the problem of unification. We conclude the chapter with the added dimension of representing and reasoning with dynamics.

2.2   Symbolic Logic

Until the 19th century, symbolic logic was primarily used for studying the foundations of mathematics. David Hilbert's program, Kurt Gödel's seminal incompleteness theorem and other such inquiries studied the arithmetic of natural numbers, and what logically follows from that arithmetic. It is only since the 1960s that John McCarthy, among others, began to see first-order logic as a formalism for capturing commonsense and language. (Although Leibniz suggested that thinking could be understood as symbol manipulation, it was only with the results of Boole and Frege that an algebraic view of logic became possible.)

2.2.1 First-Order Logic

First-order logic is powerful, and widely studied. In this book, we will assume the full language, and to contrast it with popular proposals in statistical relational learning, let us start with an often used example from that community involving smoking within social networks. We will then discuss variations admitted by propositional and first-order logic, whilst also briefly demonstrating their features. As will become clear, virtually every formalism from that community does not support arbitrary first-order logic. There is one good reason for this: first-order logic is only semi-decidable, and computing entailments is not tractable. However, since our interest in this work is to focus on the generality first, so that we have all the representational flexibility we need, and computation second, this will be a point of distinction.

Let us imagine two individuals a (for Alice) and b (for Bob) such that a is a smoker, and the two are friends. Using $S(x)$ as a predicate to mean x is a smoker and $F(x, y)$ to mean x is a friend of y, we have:

$$S(a) \wedge F(a, b)$$

There are two propositional assertions here, connected using a conjunction, to say both are true in the knowledge base (KB). Suppose we found that smokers influence those in their social networks to also smoke, in which case, we would wish to add $S(b)$ to the KB. However, rather than working this out by hand, we might wish to express this observation using an implication rule:

$$S(a) \wedge F(a, b) \supset S(b)$$

(Note that the conjunction is given precedence to the implication, so this should be read as saying $(S(a) \wedge F(a, b))$ implies $S(b)$.) Such rules involve only constants, and as such, we could instead use a rule of the below sort using abstract symbols in the KB:

$$p \wedge q \supset r$$

Keep in mind that we would, however, need to introduce such abstract symbols and rules for every possible instance of the $S(x)$ predicate. With quantifiers, we will be able to say that the smokers rule applies to every individual:

$$\forall x, y. \, S(x) \wedge F(x, y) \supset S(y)$$

The quantification is over a "domain of discourse", which defines the constants to which the expression following the quantification apply to. Suppose the domain also contained animals and plants, we might using the predicate $H(x)$ to mean x is a human to say:

$$\forall x, y. \, [H(x) \land H(y) \supset (S(x) \land F(x, y) \supset S(y))]$$

Naturally, the KB is assumed to also contain "facts", such as $S(a)$ and $F(a, b)$, in addition to such quantified formulas, and these constitute explicit knowledge. By applying these facts to the rules, we can derive $S(b)$, which is implicit, in the sense of the terminology used earlier.

Facts are meant to be ground atoms such as $S(a)$, and so, in some cases, the truth value for every instance of $S(x)$ may be determined from explicit or implicit knowledge. Such a representation corresponds to a database, and denotes the case of "complete knowledge." Incompleteness is usually expressed using disjunctions and existential quantifiers. For example, assuming a third constant c for Carol:

$$(S(a) \lor S(c)) \land \neg(S(a) \land S(c))$$

says that either Alice or Carol is a smoker. The first conjunct $(S(a) \lor S(c))$, in fact, allows both to smokers, but the second conjunct ensures that $S(a)$ and $S(c)$ cannot both be true.

The use of an existential can declare the existence of an individual with a certain property but not who that might be. The following sentence, for example, says that there is some friend of Alice, but it is not Carol:

$$(\exists x. \, F(x, b)) \land \neg F(c, b)$$

Suppose we further stipulated that one cannot be the friend of oneself:

$$\forall x. \, \neg F(x, x)$$

Of course, if our domain was finite and consisted only of the individuals discussed so far $\{a, b, c\}$, then $F(a, b)$ would follow from the above two sentences. But if the domain is not finite, infinitely many candidates are possible friends of Bob, but we are not certain of any one of them.

2.2.2 Infinite Domains

With infinite domains, in fact, other types of incomplete knowledge can also be expressed. For example, when quantifiers range over an infinite set, the following sentence:

$$\forall x. \, x \neq a \supset S(x)$$

says that there are infinitely many individuals other than Alice who are smokers, while also leaving open whether Alice is a smoker. So it models unknown atoms. To model unknown values, we might have:

$$\forall x(S(x) \supset eyecolor(x) \neq green)$$

which says that in our population, the eye color of smokers is anything but green.

Finally, if we do not specify the set of humans in advance, a sentence of the below sort can mean that smoking is true of all humans, but the set of (possibly infinite) humans can vary depending on the world:

$$\forall x (Human(x) \supset S(x))$$

Although there are a wide range of formalisms in statistical relational learning, to a large degree, constants are fixed in advance and are finite, functions are rarely allowed, and often the interpretation of predicates is fixed. Since they allow the decoration of atoms with probabilities, it allows some flexibility in terms of the interpretation of predicates, albeit defined over the finite set of constants.

In contrast, the full first-order language, especially over a possibly infinite domain, as shown above, can allow for unknown sets of atoms, their possible values, and varying sets of objects (e.g., when the set of humans is left open). This is obviously very powerful, making standard first-order logic a rich and general-purpose language for knowledge representation.

Keeping the domain undefined and not fixed in advance is often referred to as "open-world" modelling, and can be contrasted to the standard *closed-world assumption* that fixes the set of constants, and in more restrictive versions, also specifies the truth value of every atom. Such an assumption is not hard to support in the general language, should we so wish it. For example, the below sentence says that Alice and Bob are both smokers, and everyone else is not a smoker:

$$\forall x [((x = a \lor x = b) \supset S(x)) \land ((x \neq a \land x \neq b) \supset \neg S(x))]$$

If we wished to further fix the set of constants, we could relativize all sentences w.r.t. an abstract category of, say, objects. So when intending to write $\forall x (\phi(x))$, we could instead write:

$$\forall x (Objects(x) \supset \phi(x)),$$

and combine that with a specification of the (presumably finite) set of objects:

$$\forall x [((x = a \lor \ldots \lor x = b) \supset Objects(x)) \land ((x \neq a \land \ldots \land x \neq b) \supset \neg Objects(x))].$$

2.3   Probabilities on Formulas

The standard apparatus to capture incomplete knowledge in symbolic logic is by means of disjunctions and existential quantifiers. Consider that a logical disjunction expresses qualitative uncertainty, but does not commit to which of the disjuncts is more likely than the other. In almost all applications of interest, uncertainty is a fact of life, and in very many cases, we would have good reasons to believe some events are more probable than others. We might even data to support the likelihood of one event over another.

For example, suppose we toss a coin and find that it lands on its head nine out ten times. We might believe it is a loaded coin, and ascribe a probability of 0.9 to observing a heads on the next toss. Suppose we read three plausible reports, two of which suggest that Alice is not a smoker, but a third one asserting that she is one. We might ascribe a probability of 2/3 to $S(a)$. However, given that a report about Alice being a smoker might have been more recent or a case where Alice accidentally admitted she occasionally smokes, perhaps we might treat it as a true but lesser known fact, and ignore the other reports. Whichever the case may be, we need a language to reason about the *degrees of belief*—i.e., subjective probabilities—of the agent.

Perhaps the simplest and most widely studied proposal for dealing with degrees of belief in a logic is based on the notion of *possible worlds*—where each world is a distinct interpretation of the underlying knowledge base—and we ascribe a probability (or more generally, a weight) to each such world. The probability of, say, $S(a)$ is obtained by summing the probabilities of all worlds where $S(a)$ is true. If we use weights, we might then consider the ratio of that sum to a *normalization factor*, which is simply the sum of the probabilities of all worlds.

It is this framework that we will use in the rest of the book. Statistical relational languages essentially deal with propositional possible worlds. (We will return to this point at the end of this section.) There are three key differences to those proposals from the logical language in this book: (a) we will entertain first-order structures as possible worlds, (b) we will allow for non-unique probability distributions, and (c) we will admit a general theory of dynamics. Note, for example, that since our knowledge bases can be any first-order theory (and further decorated with probabilities), we will also need to deal with countably infinite worlds. To deal with arbitrary histories of actions, these worlds are trees, where nodes capture what is true about the world after a corresponding sequence of actions have happened.

The subsequent chapters will introduce our logical framework, but in the meantime, let us consider a few examples with a probabilistic logical language. Those examples involving propositions and single probability distributions can be captured with statistical relational learning languages, whereas the first-order ones over infinite domains and imprecise probabilities cannot.

2.3.1   Probabilities on Atoms

Suppose $S(a)$ is believed with a probability of 1. What is the probability that Bob is also a smoker? Consider that $S(a)$ being believed with a probability of 1 means that in every possible model of the KB, $S(a)$ holds; in other words, $S(a)$ is entailed by KB. Given that, what can we say about $S(b)$? Clearly, this would very well depend on the possible worlds. If it was also the case that $S(b)$ was entailed by the KB, it would be accorded a probability of 1 as well. However, if some models of the KB did not satisfy $S(b)$, then by summing the probability of all such worlds, we would obtain the probability r of $\neg S(b)$. And so, the probability of $S(b)$ would be $(1 - r)$. Put simply, if the worlds all had equal weights and $S(b)$ was false in r out of n worlds, then the probability of $S(b)$ would be $(n - r)/n$. From a logical viewpoint, the extension of the predicate S is different in every world, and it so happens a is in every such extension and b is only in some.

Suppose, based on collected data, we have a "noisy" rule of the sort in the KB:

$$4/5 \quad (S(a) \wedge F(a, b)) \supset S(b)$$

If the worlds all had equal weights (for ease of explanation), this is basically saying that in four-fifths of the worlds, Alice influences her friend to smoke.

To reason about the rule further, consider two scenarios with different probabilities for Alice being a smoker. Suppose $\{S(a), F(a, b)\}$ is believed with probability 1. Owing to the noisy rule $S(b)$ must be true in four-fifths of the worlds, and so we would obtain a probability of 4/5 for $S(b)$. On the other hand, suppose $F(a, b)$ is believed with probability 1 but $S(a)$ only with a probability of 2/3. Assume equal weights to worlds for ease of explanation. Then, in two-thirds of the worlds, $S(a)$ holds, and in four-fifths of those, the noisy rule would require that $S(b)$ holds in those. Therefore, the probability of $S(b)$ would be $2/3 \times 4/5 = 8/15$.

Let us consider a simple example where a unique distribution would not arise. Suppose $S(a) \vee S(c)$ is believed with probability 1. In this case, no specification is provided for the individual probabilities of $S(a)$ and $S(c)$. It is not permissible to accord a probability of 0 to both $S(a)$ and $S(c)$. Provided that requirement is satisfied, many possibilities exists. Most statistical relational languages will either disallow such specifications or make assumptions about how the distribution should be defined for such disjunctions. (Notable exceptions allowing for imprecise probabilities do exist, but often not in a first-order formalism.)

Suppose we are interested in the probability of $S(b)$ given the noisy smokers rule above, and a similar instance as a result of $S(c)$:

$$4/5 \quad (S(c) \wedge F(c, b)) \supset S(b)$$

Suppose $\forall x, y(F(x, y))$ is believed with probability 1, that is, every individual is a friend of everybody else (including himself). It is not hard to see the probability of $S(b)$ should be 4/5. This is because every world should satisfy $S(a)$ or $S(c)$ or both. Owing to the universal about everyone being friends, $F(a, c)$ and $F(b, c)$ both hold. Consequently, in four-fifths of the worlds (assuming equal weights for simplicity), the smokers rule would mean $S(b)$ must hold. Therefore, its probability is also 4/5.

Suppose we instead said $S(a) \vee S(c)$ only with a probability of 1/5. Assuming equal weights, in one-fifth of the worlds, $S(a)$ or $S(c)$ or both. In four-fifths of those, the smokers rule would mean $S(b)$ must hold. Therefore, the probability of $S(b)$ would be $4/5 \times 1/5 = 4/25$.

2.3.2   Probabilities on Quantified Formulas

Our examples from the previous section would work just as well (and more succinctly) with a noisy quantified rule of the sort:

$$4/5 \quad \forall x, y(S(x) \wedge F(x, y) \supset S(y))$$

This is saying that in four-fifths of the worlds (assuming equal weights to worlds), friends of smokers are smokers too. Suppose we were certain Alice and Bob were friends, and we ascribed a probability of $2/3$ to the atom $S(a)$. Then in two-thirds of the worlds $S(a)$ holds, and in four-fifths of these, the universal together with the certain knowledge that $F(a, b)$ means $S(b)$ holds. The probability of $S(b)$ is then $2/3 \times 4/5 = 8/15$.

However, one might find the contrast between the reading of this rule and the statistical observation motivating that rule striking. Presumably, we would like to say in 80% of the population, friends of smokers are smokers too. This is statistical information, of course, but the way the rule is written requires that in four-fifths of the worlds, the rule must hold. In other words, if in every world, there is some smoker whose friend is not a smoker, then the rule clearly holds in no world, and so its probability should be 0.

This is an early observation in the field of probabilistic logics and pertains to the fact that the possible-worlds model with probabilities over worlds is used to define degrees of belief. It is not appropriate for statistical information. Given that we know the smokers universal not to be true in the real world, it would be useful to represent statistical information as well. This requires a different formal setup—a so-called *random-worlds* semantics—where probabilities can be ascribed to domain elements. This would allow us to reason about the probability of a randomly chosen person is a smoker, or that of two randomly chosen individuals being friends, and so on.

In most statistical relational languages, this intended statistical interpretation is simply ignored, and noisy quantified rules of the above sort are commonplace, all interpreted using the possible-worlds model.

In this book, we expect to use quantified formulas obviously, given the first-order language, but these are always believed with probability of 1. For example, as we shall see, the axiomatization of the domain—which will involve quantification over actions—will be believed in all possible worlds.

While we are not opposed to assigning less than 1 probability to quantified formulas in the knowledge base, and there is certainly no technical restriction that requires this, it should be clear that this is really saying that a certain proportion of the worlds satisfies the quantified formula, and not that a certain proportion of the individuals apply to the formula. Likewise, probabilities on existentially quantified formulas is to be understood the same way. For example:

$$1/5 \quad \exists x(S(x))$$

says that in one-fifths of the worlds (assuming equal weights on worlds), there is someone who is a smoker. It is *not* saying: one of five individuals is a smoker, or that there is a 20% chance a randomly chosen individual is a smoker.

Both of these require a random-worlds semantics, and integrating that with our possible-worlds semantics would be an exciting line of research.

It is interesting to note that we will nonetheless discuss how beliefs can be integrated with noisy sensors and effectors, whose error profiles are obtained by statistics.

2.3.3 Essentially Propositional Languages

We previously mentioned that statistical relational languages are essentially propositional. There is a fair bit of caveat to be added to that remark.

While many such languages are defined by simply providing a relational template to standard propositional formalisms such as Bayesian networks, there is also a good deal of work extending that limitation in various ways. We review three popular proposals to elaborate. Although from these discussions, it will become clear that the logical language of the book is considerably more general, we also believe they are attractive candidates for initial knowledge bases in our formal setup. Indeed, our reasoning results establish that reasoning about dynamics can reduce to reasoning about the initial KB, and so we may choose to use such languages at that stage for computational reasons.

In Markov logic networks, a relational probabilistic model developed at the University of Washington, every first-order formula is interpreted over a finite domain, and so existentially quantified formulas can be reduced to a large disjunctive formula. However, the notion of logical entailment is relaxed, in the sense that the weights on the possible worlds are adjusted depending on which formulas from the KB are satisfied. This means that not every world is required to satisfy the KB. So when computing the probability of a formula given a KB, we will also consider worlds that are not models of the KB. This clearly deviates from the semantics for degrees of belief discussed earlier. Such a relaxed semantics is motivated by the need to allow multiple KBs (as obtained from, say, different sources), possibly with conflicting information.

In ProbLog, the probabilistic logic programming language developed at KU Leuven, probabilities can decorate atoms in logic programs. This essentially amounts to probabilities being accorded to possible valuations to ground atoms. Such programs can contain cycles, and so, in principle, could lead to infinitely many atoms. But in practise, the grounding of the program is only considered w.r.t. a ground query atom. This means that there is a semantically well-defined conversion of ProbLog programs to "weighted propositional theories," which are propositional formulas with non-negative weights on atoms.

In fact, ProbLog, probabilistic databases, Markov logic networks, Bayesian networks, factor graphs, and a wide range of statistical relational learning formalisms can be understood in terms of a computational framework called *weighted model counting*. Given a logical language with n propositions, and thus, 2^n worlds $\mathcal{M} = \{M_1, \ldots, M_{2^n}\}$, consider a weight function μ mapping literals to non-negative reals. By extension, let the weight of a world $\mu(M)$ be defined as the product of the weights of the literals that hold at M. Then the weighted model count of a propositional formula ϕ is defined as:

$$WMC(\phi) \doteq \sum_{\{M \in \mathcal{M} | M \models \phi\}} \mu(M) \Big/ \sum_{\{M \in \mathcal{M}\}} \mu(M)$$

A Bayesian network may be encoded as a propositional knowledge base, such that the conditional dependences get expressed as implications; and, in Markov logic networks, propositions appearing in a clause would mean that they correspond to the dependent random variables in an undirected graphical model. The probability function of the nodes may be further expressed in terms of a weight function of the above sort, in which case, the probability of some query q (a propositional formula) is obtained using:

$$WMC(KB \wedge q) \Big/ WMC(KB)$$

In other words, relative to the models of KB, we compute the weights of those worlds additionally satisfying q.

It is not hard to see that this is essentially a propositional analogue to the semantics of degrees of belief described earlier. We will also include an explicit belief operator in our language allowing us to reason about meta-beliefs in this book.

In the probabilistic programming language BLOG from UC Berkeley, function symbols are allowed in the language over an infinite domain. Thus, it supports open-world modelling. However, for computational reasons, various restrictions are placed both for the language and its implementation. Well-defined programs, for example, restrict the use of logical connectives. When computing probabilities, it avoids instantiating infinitely many terms by sampling values for function symbols. When it comes to termination, moreover, a number of structural conditions need to hold, including one that the quantifiers over formulas can only range over finite sets.

As mentioned above, these languages are carefully designed to allow for logical reasoning whilst remaining computationally feasible. While they are clearly less restrictive than the general-purpose logical framework we consider, exploring the use of such probabilistic logical languages as an initial KB is a compelling research direction.

2.3.4 Actions

In addition to the first-order and imprecise probability aspects, the third dimension to our language is that it is built on a general theory of actions. We should remark that many statistical relational languages have been extended to handle domain dynamics, and some even to support noisy acting, sensing and planning. (A comprehensive discussions on such proposals will be considered in the subsequent chapters.) Like with our arguments in the previous sections, the goal of our work is study a general model of actions, where we can better position philosophical concerns such as the frame, qualification and ramification problems, but also allow for fine-grained temporal analysis of actions and events. A general theory of actions allows us to write formulas expressing involved statements about past occurrences, hypotheticals and counterfactuals. Such capabilities are not present in most statistical relational learning and planning frameworks, which might be designed to either track beliefs after actions, or synthesize plans to enable goals. (Temporal logics are notable exceptions, but they are closer to "action logics" such as the situation calculus, as we shall see.)

Of course, when the application context is known, a pragmatic choice for the language needs to be made. In that case, much like the need for a tractable initial knowledge base, there is also the need for an effective belief tracking and planning framework that operates within a reasoning module. As hinted above, we ignore such considerations here.

In the next chapter, we will formally introduce the logical language for reasoning about actions, and then discuss how to extend that to knowledge and degrees of belief. The resulting language, as we shall see, will allow for first-order probabilistic assertions over a rich theory of actions. Perhaps for readers unfamiliar with the situation calculus, the semantics of belief might not make it very obvious that, as suggested by our discussion above, we are defining a probability measure over possible worlds. Moreover, in the situation calculus, these worlds are trees. Each node in such a tree corresponds to a situation, and interprets first-order formulas about the current history. The penultimate section provides a modal reconstruction of our language, and this semantic perspective of measures-on-possible-trees is brought up in arguably a more crisp and clean fashion.

2.4   Notes

For an introduction on first-order logic, see classic textbooks such as [69, 206]. Closed-world and open-world semantics is discussed in [190], among others, and logic programming in [128].

For the social network example on smokers influencing their friends, see [187, 192]. In that regard, [192] is an authoritative account on Markov logic networks, and [187] introduces ProbLog is significant detail. Technical works that develop the semantics and computational machinery of ProbLog can be found in [76, 188].[1] See [89, 90] for some other relational probabilistic languages from the statistical relational learning community.

---

[1] Logic programs make the CWA allowing them to capture inductive definitions [66], which are not expressible in first-order logic. But they are expressible in second-order logic, as shown for the situation calculus [67, 213].

Probabilistic databases are comprehensively studied in [210], factor graphs in [132], and *weighted model counting* in [6, 43, 43, 194, 225]. BLOG is introduced in [169]. Probabilistic relational models for dynamic domains from this community are considered in [141, 175, 176, 196], among others. For a treatment on imprecise probabilities, see [52].

In-depth technical analyses on general first-order logics of probability is considered in [5, 100], including the random-worlds model.[2]

---

[2] We remark that there is work on languages integrating degrees of belief and the random-worlds semantics; see, for example [7, 100]. However, further integrating that with the advances on reasoning about actions considered in this book is yet to be pursued.

From Predicate Calculus to the Situation Calculus 3

You can prove anything you want by coldly logical reason—if you pick the proper postulates.

– Isaac Asimov, I, Robot

Our account is formulated in the language of the situation calculus, the most established special-purpose knowledge representation formalism for reasoning about dynamical systems. Originally postulated by McCarthy, and later revised by Reiter, it is a dialect of first-order logic with distinguished sorts. So we begin with a brief recap of standard first-order and second-order logic, before turning to the situation calculus.

We reiterate that although concepts are explored formally in this book, we attempt to motivate the formalism and results using intuition, rather than developing proofs or other such formal exercises. In other words, familiarity with first-order logic would make the reading easier, but key ideas could be grasped with a cursory understanding too.

3.1 Predicate Calculus

Formal logic is characterized by a syntax and a semantics. The syntax defines the symbols, terms and formulas that can be constructed, and the semantics provides a means to interpret these symbols. We only go over the essentials.

The syntax assumes a logical signature consisting of a set of predicate symbols of every arity, a set of function symbols of every arity, equality, logical variables, and standard connectives ($\wedge$, $\vee$, $\neg$, $\forall$). Parentheses are also allowed for grouping formulas together, to improve readability. We use $P \supset Q$ to abbreviate $\neg P \vee Q$ and $P \equiv Q$ to mean $(P \supset Q) \wedge (Q \supset P)$. We use $\exists x P(x)$ to abbreviate $\neg \forall x \neg P(x)$.

© The Author(s), under exclusive license to Springer Nature Switzerland AG 2023
V. Belle, *Toward Robots That Reason: Logic, Probability & Causal Laws*,
Synthesis Lectures on Artificial Intelligence and Machine Learning,
https://doi.org/10.1007/978-3-031-21003-7_3

Terms are obtained from variables and functions, atoms from predicates taking terms as arguments, and formulas by composing atoms over connectives. A sentence is a formula without free variables.

To interpret atoms and formulas, and accord them a truth value, we define first-order models, aka Tarskian structures. A *model M* is defined over a universe $|M|$, a non-empty set of objects, which serves as the domain of discourse. For every k-ary predicate P, $P^M \subseteq |M|^k$. Analogously, for every k-ary function f, f^M is a mapping from $|M|^k$ to $|M|$. In particular, when the function is nullary, $f^M \in |M|$, that is, the constant symbol is an element of the universe.

In other words, a model specifies the set of elements over which we quantify. Interpreting a predicate corresponds to a set of k-tuples, and interpreting a function corresponds to a mapping from such k-tuples (understood as arguments of the function symbol) to an element of the universe.

Given a formula ϕ, we define truth w.r.t. a structure M inductively. To do this, we need a mapping function for interpreting variables—*a variable map*—as well as a way to recursively identify the elements of the domain for every term in a formula. Recall that a term can be obtained from function symbols applied to any depth, such as $f(g(h(c)))$ and $h(g(g(x)))$. We omit the details, but for example, given the atom $P(c_1, \dots, c_k)$, where the arguments are constants, we have:

- $M \models P(c_1, \dots, c_k)$ iff $\langle c_1^M, \dots, c_k^M \rangle \in P^M$.

(We abbreviate "if and only if" as "iff.") That is, we identify the elements of the domain for each of the constants, and then check whether the tuple of these elements is in the set P^M, which is the interpretation of P in the model M. (It is also referred to as the *extension* of P.) With atoms sorted, we inductively define:

- $M \models \neg\phi$ iff it is not the case that $M \models \phi$ (often written $M \not\models \phi$);
- $M \models \phi \wedge \psi$ iff $M \models \phi$ and $M \models \psi$; and
- $M \models \forall x \phi(x)$, where x is the only free variable in ϕ, iff for every $d \in |M|$, $M \models \phi(d)$. Here, $\phi(d)$ is the formula $\phi(x)$ but with x replaced by the element d of the universe.

We then say a sentence ϕ is *satisfiable* iff there is a model M such that ϕ is true in the model, that is, $M \models \phi$. The sentence is *valid* iff ϕ is true in every model. For a set of sentences $\mathcal{D}$, we write $\mathcal{D} \models \phi$ (read: $\mathcal{D}$ *entails* ϕ) to mean that for any model M, if $M \models \alpha$ for every $\alpha \in \mathcal{D}$, then $M \models \phi$.

It is easy to extend the above language to include sorts, which can capture different types of entities in the universe. For example, we may want to distinguish numbers from animals, and perhaps even distinguish integers from reals, with the understanding that when we use quantifiers, we expect the domain of quantification to only apply to the appropriate sort. For example, we may wish to write $\forall x (Even(x) \supset \phi(x))$, with the intention that the x only ranges over integers, and the argument to the predicate only takes terms of the integer sort. To do this, we would need to assume quantifiers for each sort, and the extension of the predicate would be interpreted over the appropriate sub-universe. It is worth noting, however, this is purely syntactic convenience, as the functionality can be replicated by simply allowing for a unary predicate $Integer(x)$, whose extension is precisely the integers in the universe. For example, the above formula could then be written as $\forall x[Integer(x) \supset (Even(x) \supset \phi(x))]$.

The situation calculus also makes use of second-order logic to define its ontology of possible worlds. Second-order logic admits quantification over predicate and function symbols. Thus, the language additionally allows predicate variables and function variables. So expressions such as $\forall P\ \phi$, where ϕ is a well-formed formula in first-order logic, is now a well-formed formula in second-order logic.

While the definition of a model is the same as before, we will need to extend the variable maps to now also interpret second-order variables. That is, where previously we just mapped variables to the universe $|M|$, we will now need to map the k-ary predicate variable P to a k-ary relation on the universe. With that, the definition of truth is given inductively as before, but with the added rule for second-order quantifiers:

- $M \models \forall P\ \phi$ iff for every k-ary relation $R \subseteq |M|^k$, $M \models \phi_R^P$. Here, ϕ_R^P is exactly like ϕ but with every occurrence of P replaced by R; in other words, for every possible extension of the k-ary predicate variable P, ϕ is true in the model.

Of course, we would need to provide a rule for function variables analogously. And, with that, validity and entailment is defined as before.

3.2 A Theory of Action

Our account is formulated in the language of the situation calculus. It is now widely studied, and was originally proposed by McCarthy, dating back to the early days of AI. It has the advantage that it is essentially a dialect of standard first-order logic, with no new model theory needed. It is a dialect because it adds some distinguished terms, and sorts for actions, objects and so-called situations, all leading to a very simple ontology. The modelling approach here is via *axioms,* that is, sentences in the language capturing the domain of interest. We reason about the domain by checking whether certain properties are entailed by those axioms. Before introducing the language formally, let us draw an intuitive picture.

3.2.1   Ontology and Assumptions

The situation calculus will allow us to express things that are true initially, but also after any sequence of actions. So a "possible world" is not just a static state of affairs, but a tree-like structure for capturing the branching nature of a dynamic system affected by the agent's actions. Imagine the world before any actions have occurred, an empty history if you will. From here, executing an action a leads to one state of affairs, but executing b leads to another. After executing one of these, we might find ourselves yet again with the possibility of executing either a or b, leading to one of possible histories $a \cdot a, a \cdot b, b \cdot a, b \cdot b$, and so on. So far, so good. However, any theory of action, not just the situation calculus, has to grapple with a number of representational issues. Addressing these issues requires us to make some modeling assumptions.

First, what does the world look like and how is the world being affected? We are supposing here that the world is captured by a set of first-order formulas. In principle, this is not that different from a database capturing a static state of affairs. But a database is equivalent to a finite set of literals, so a set of first-order formulas is significantly more general. Likewise, modelling and verification frameworks in computer science formalize system properties as a set of propositions, or similar, where, fundamentally, a commitment is being made in terms of how much of the world is captured and at which granularity. (Arguably, a putative autonomous agent might need to understand the world at different levels of granularity, encompassing knowledge about everything from gravity to subatomic interactions. While such a commonsense theory is still to emerge, first-order logic and its probabilistic extensions offer a promising strategy for modelling such a theory to date.)

Next, on performing actions, the truth values of these formulas may change. For example, dropping a fragile object leaves it broken after the action. Moreover, the world is only changing as a result of such *named actions*. This does not mean that actions need to serve as a proxy for time. It simply means that properties of the world are affected only by actions. Fortunately, we do not have to insist that all named actions are only executable by a distinct putative agent. Modelling nature and multiple agents are permitted.

Of course, there will also be properties that do not change at all, such as a bird being a penguin rather than an ostrich, and the object being fragile. We could contrast such *rigid* properties with *fluent* properties, such as the breaking of objects, which change with actions. Thus, the language will include both rigid and fluent predicates for capturing unchanging and changing relations, and both rigid and fluent function symbols for capturing unchanging and changing functions. The biological father of an individual is unchanging, but the citizenship of an individual may change, for example. (The biological father may not be known to the agent, or otherwise, believed falsely. Capturing these distinctions is the focus of the next chapter.)

Second, what conditions must be true for executing an action? Consider the action of picking up a box. Observe that for executing the action, we need to make some assumptions about what is true in the current state. The agent presumably is not holding anything else, unless the box is light and it can be picked up using only a single hand. The box should also not so heavy that it cannot be picked up even if both hands are free. The box needs to be next to the agent, say, within grasping distance. But beyond these simple conditions, we can elaborate on a large number of hypotheticals, such as: the agent is alive, the box is not glued to the floor, there is sufficient oxygen for the agent to breathe and move, the agent's hands are not severed, and so on. Clearly, it would be impossible to specify all such *qualifications* for being able to perform the action. This is referred to as the *qualification problem*. The assumption we make is that there is a small set of conditions—a first-order sentence β—such that an action can be executed iff β. The interpretation is that β is both necessary and sufficient for executing the action, but we are ignoring minor qualifications (such as being glued to the floor).

Third, how do we capture all the properties affected by, but also those not affected by the execution of an action? On performing the action of picking up the box, we expect the agent to now be holding the box, as well as the position of the box to have changed: it is no longer on the floor. But, we do not expect the color of the box to have changed. (Unless, of course, the box contained paint and in the process of it being picked up some of the paint split and colored the outside of the box.) We also do not expect the temperature of the room to change. There are potentially a large number of things that remain unchanged. We need a succinct way for writing down action invariants, leading to the so-called *frame problem*. McCarthy's situation calculus became workable and practical with Reiter's monotonic solution to the frame problem. As McDermott put it, the solution "breathed new life into the situation calculus". It consists of simply specifying a first-order formula γ, one for every fluent, with the understanding that the conditions under which actions make the fluent true together with the conditions that making the fluent false are expressed in γ. These are referred to as successor state axioms, introduced shortly. Implicit in the form of these axioms is the *causal completeness assumption*, which states that *all* such conditions are in γ.

Actions might also have indirect effects, including secondary and tertiary effects (toggling a switch will turn the light on, but toggling every switch will not only turn all lights on but also blow the fuse), and delayed effects (turning on the faucet in a tub with the valve closed will fill the tub, until it spills over and wets the floor). Modelling such actions is possible in the language, but requires additional machinery. In fact, understanding ramifications, modelling time, concurrency, durative effects, and procedures are amenable to formalization (to a reasonable degree) in the situation calculus. We only discuss procedures here, and follow Reiter in ignoring ramifications. The interplay between time and actions is interesting, but also not dealt with further here. (Time can either be modeled explicitly using a new sort, or implicitly linked to the execution of actions.)

It is worth noting that none of the assumptions we make about the changing world are controversial from the standpoint of practical AI, at least to the extent of current formalisms and its applications. Automated planning formalisms specify preconditions, along with positive and negative effects as a set of literals, so its a special case of the first-order language we have. Decision-theoretic and reinforcement learning approaches are based on state-based abstractions, specifying the set of actions available at states and the successor states obtained on executing one of the available actions. These too can be seen as a propositional reworking of the more general model afforded by the situation calculus. The key feature of a formal language is that these issues are made transparent, because we need to explicitly specify the properties of the world, and how they change. Given our aims in this book, this is an important exercise.

3.2.2 The Language

The formalism is best understood by arranging the world in terms of three kinds of enti-ties: *situations*, *actions* and *objects*. Situations represent "snapshots," and can be viewed as possible histories. An initial situation corresponds to the way the world can be prior to the occurrence of actions. The result of doing an action, then, leads to a successor (non-initial) situation. Dynamic worlds change the properties of objects, which are captured using pred-icates and functions whose last argument is always a situation, called *fluents*. In contrast, *rigids* do not have a situation argument.

Formally, the language $\mathcal{L}$ of the situation calculus is a many-sorted dialect of predicate calculus, with sorts for *actions*, *situations* and *objects* (for everything else). In full length, let $\mathcal{L}$ include:

- logical connectives $\neg, \forall, \wedge, =$, with other connectives such as $\supset$ understood for their usual abbreviations;
- an infinite supply of *variables* of each sort;
- an infinite supply of *constant* symbols of the sort object;
- for each $k \geq 1$, *object function* symbols $g_1, g_2, \ldots$ of type $(action \cup object)^k \to object$;
- for each $k \geq 0$, *action function* symbols $A_1, A_2, \ldots$ of type $(action \cup object)^k \to action$;
- a special *situation function* symbol *do*: $action \times situation \to situation$;
- a special predicate symbol *Poss*: $action \times situation$;
- for each $k \geq 0$, *fluent predicate* symbols $P_1, P_2, \ldots$ of type $(action \cup object)^k \times situation$;
- for each $k \geq 0$, *fluent function* symbols $f_1, f_2, \ldots$ of type $(action \cup object)^k \times situation \to object$; and
- a special constant S_0 to represent the actual initial situation.

To reiterate, apart from some syntactic particulars, the logical basis for the situation calculus is the regular (many-sorted) predicate calculus. Terms and well-formed formulas are defined inductively, as usual, respecting sorts. (Note that, by extension, we will subsequently only need to introduce a few more distinguished predicates for modeling knowledge, sensing and nondeterminism.)

To get familiar with the language, consider saying that initially the box c is on the floor, not held by the agent, and that all boxes are colored red. We may write:

$$OnFloor(c, S_0) \wedge IsBox(c) \wedge (\forall x (IsBox(x) \supset Color(x, red, S_0))) \wedge \neg Holding(c, S_0).$$

Here, we are treating an object being a box as an unchangeable property, and hence $IsBox(x)$ is a rigid, and does not have a situation term. In contrast, being on the floor and the current color are properties that can change. Hence, they are fluents. Moreover, fluents have a situation term as the last argument, and in particular, by letting this argument be S_0, we have specified how things are initially.

Dynamic worlds are enabled by performing actions, and in the language, this is realized using the *do* function. That is, the result of doing an action a at situation s is the situation $do(a, s)$. Fluents, which take situations as arguments, may then have different values at different situations, thereby capturing *changing properties* of the world. For example, the agent is not holding things initially, so after picking up the box, we may say:

$$Holding(c, do(pickup(c), S_0)) \wedge Color(c, red, do(pickup(c), S_0)).$$

That is, when considering the successor situation $do(pickup(c), S_0)$ obtained by performing $pickup(c)$ at S_0, the agent is now holding the box. But also, of course, we do not expect the color of the box to change, so it should also hold that the box is still red. Analogously, if the agent paints the box blue whilst holding the box, we get:

$$Holding(c, s') \wedge Color(c, blue, s'),$$

where $s' = do(paint(c, blue), do(pickup(c), S_0))$. Thus far, we have simply used the original language from McCarthy. The use of basic action theories, which includes successor state axioms incorporating Reiter's solution to the frame problem are discussed below.

Before getting into these, let us introduce some notation. We follow some conventions in the ways we use Latin and Greek alphabets: a for both terms and variables of the action sort (the context would make this clear); s for terms and variables of the situation sort (the context would also make this clear); and finally, x, u, v, z, n, and y to range over variables of the object sort. We let ϕ and ψ range over formulas, and Σ over sets of formulas. (These may be further decorated using superscripts or subscripts.)

We sometimes suppress the situation term in a formula ϕ, or use a distinguished variable *now*, to denote the current situation. Either way, we let $\phi[t]$ denote the formula with the restored situation term t. For example, given a fluent h, the formula $h(S_0) = 5$ could be written as $(h(now) = 5)[S_0]$. It could equivalently be written as $(h = 5)[S_0]$, where it is obvious from the context that the situation term applies to the fluent h.

Let α denote sequences of action terms or variables, and we freely use this with do, that is, if $\alpha = [a_1, \ldots, a_n]$ then $do(\alpha, s)$ stands for $do(a_n, do(\ldots, do(a_1, s) \ldots))$.

In addition to the usual IF-THEN-ELSE notation (examples will follow in subsequent chapters), we will also use the "case" notation with curly braces as a convenient abbreviation for a logical formula:

$$z = \begin{cases} t_1 & \text{if } \psi \\ t_2 & \text{otherwise} \end{cases} \doteq (\psi \supset z = t_1) \wedge (\neg\psi \supset z = t_2).$$

Finally, for convenience, we often introduce formula and term abbreviations that are meant to expand as $\mathcal{L}$-formulas. For example, we might introduce a new formula A by $A \doteq \phi$, where $\phi \in \mathcal{L}$. Then any expression $E(A)$ containing A is assumed to mean $E(\phi)$. Analogously, if we introduce a new term t by $t = u \doteq \phi(u)$ then any expression $E(t)$ is assumed to mean $\exists u(E(u) \wedge \phi(u))$.

3.2.3 Basic Action Theories

Domains are modeled in the situation calculus as *axioms*. A set of $\mathcal{L}$-sentences specify the actions available, what they depend on, and the ways they affect the world. Specifically, these axioms are given in the form of a *basic action theory*.

In general, a basic action theory $\mathcal{D}$ is a set of sentences consisting of (free variables understood as universally quantified from the outside):

- an *initial theory* $\mathcal{D}_0$ that describes what is true initially;
- *precondition axioms*, of the form $Poss(A(\vec{x}, s)) \equiv \beta(\vec{x}, s)$ that describe the conditions under which actions are executable;
- *successor state axioms*, of the form $f(\vec{x}, do(a, s)) = u \equiv \gamma_f(u, \vec{x}, s)$, that describe the changes to fluent values after doing actions; and
- domain-agnostic *foundational axioms*.

The formulation of successor state axioms, in particular, incorporates Reiter's monotonic solution to the frame problem.

Foundational axioms ensure that the space of situations forms a tree-like structure rooted in S_0, where the nodes of the tree are the situations and the edges represent actions connecting situations with their successors, illustrated in Fig. 3.1. We omit the details here, but note that it includes a second-order axiom:

$$\forall P \, [P(S_0) \wedge \forall s, a. \, [P(s) \supset P(do(a, s))]] \supset \forall s. \, P(s).$$

This is similar to the induction axiom for the natural numbers, and defines the set of all situations to be precisely those that are obtained from S_0 and sequences of actions. In fact, the foundational axioms are all together based on Peano's axioms for arithmetic, and declare that S_0 has no predecessor, if the successor of s and s' are the same, then $s = s'$, and so on.

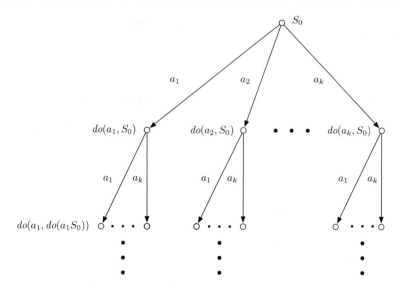

Fig. 3.1 A tree of situations

An agent reasons about actions by means of entailments of a basic action theory $\mathcal{D}$. A fundamental task in reasoning about action is that of *projection*, where we test which properties hold after actions. Formally, suppose ϕ is a situation-suppressed formula or uses the special symbol *now*. Given a sequence of actions a_1 through a_n, we are often interested in asking whether ϕ holds after these:

$$\mathcal{D} \models \phi[do([a_1, \ldots, a_n], S_0)]?$$

This concludes our review of the basic features of the language.

Let us do an example of a basic action theory. We will also briefly review two fundamental computational mechanisms for dealing with projection: *regression* and *progression*. We then review the high-level action programming language GOLOG.

In the subsequent chapters, we will discuss how the formalism is first extended for knowledge, and then, degrees of belief against discrete probability distributions and beyond. Regression, progression and high-level programming will have corresponding analogues in the presence of probabilities.

3.2.4   Axiomatization: A One-Dimensional Robot

Imagine a robot moving towards a wall as shown in Fig. 3.2, and a certain distance h from it along the horizontal axis. Moving forwards brings the robot closer to the wall, and conversely, moving backwards gets the robot away from the wall. Let us suppose moving forward is only

Fig. 3.2 A one-dimensional
robot attempting to move
towards the wall

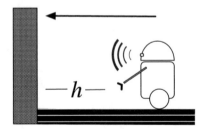

possible if the robot is not already at the wall. Then we may have the following precondition
axiom:

$$Poss(a, s) \equiv (a = move \wedge h \neq 0) \vee (a = reverse \wedge true).$$

That is, moving backwards can always be executed.

We specify the following successor state axiom for h:

$$h(do(a, s)) = u \equiv a = move \wedge h(s) = u + 1 \vee$$
$$a = reverse \wedge h(s) = u - 1 \vee$$
$$a \neq move \wedge a \neq reverse \wedge h(s) = u.$$

This is incorporating Reiter's solution for the frame problem in postulating that *move* and
reverse are the only actions affecting h, otherwise its value does not change from s to
$do(a, s)$. Moreover, for h to have a value of $u + 1$ at s would mean it has the value u at
$do(move, s)$. In other words, the distance between the robot and the wall has reduced by a
unit. Analogously, for moving backwards with *reverse*.

Example 3.1 Suppose the initial theory is

$$(h = 5)[S_0].$$

Suppose $\mathcal{D}$ is the union of this formula with the foundational axioms, and the above precondition and successor state axiom. We would have the following entailments of $\mathcal{D}$:

1. $h(do(move, S_0)) = 4$.
2. $h(do(move, do(move, S_0))) = 3$.
3. $h(do(reverse, S_0)) = 6$.
4. $h(do(reverse, do(move, S_0))) = h(do(move, do(reverse, S_0))) = 5$.

That is, moving forward once means the robot is only 4 units away from the wall, and a
second move means now it is 3 units away. Had the robot instead moved backwards at S_0,
the robot would now be 6 units away. Getting the robot to move forward and backwards, or
backwards and forwards leads to h having the same value it was taking initially.

Note that the initial theory need not determine h's value exactly. Indeed, if the initial theory was instead $(h = 5 \vee h = 6)[S_0]$, the corresponding basic action theory would entail:

$$(h = 4 \vee h = 5)[do(move, S_0)].$$

That is, regardless of which value h takes, a forward action reduces its value by 1. Semantically, the formula $(h = 5 \vee h = 6)[S_0]$ leads to some models of the action theory where $(h = 5)$ initially, and others where $(h = 6)$ initially. In the former, $(h = 4)$ after moving forward, but in the latter, $(h = 5)$ after that action. Since we are looking at the entailments of $\mathcal{D}$, which by construction considers all possible models, we find that $(h = 4 \vee h = 5)[do(move, S_0)]$ is true in all of them.

By extension, $\mathcal{D}$ also entails $h \geq 2$, $h \neq 0$, $h \geq 4$ and $h \leq 5$ in the situation $do(move, S_0)$. Likewise, if the initially theory simply said $(h < 5)[S_0]$, a forward action would now mean $h < 4$ is true in the situation $do(move, S_0)$.

3.2.5   Regression and Progression

Reasoning about the properties of a basic action theory amounts to solving the projection problem. However, the action theory includes among other things, sentences that quantify over actions and situations in the form of precondition and successor state axioms. Moreover, the second-order induction axiom would mean we need to consider second-order logic theorem proving. None of this makes projection a straightforward computational problem to solve.

What would be desirable? The initial theory is, of course, a classical first-order theory; reasoning about its entailments is well-understood. Could projection be solved by appealing somehow only to the initial theory? Fortunately, the answer is yes.

There are two main solutions to projection: *regression* and *progression*. Both of these have proven enormously useful for the design of logical agents, essentially paving the way for *cognitive robotics*. Roughly, regression reduces a query about the future to a query about what holds initially. Progression, on the other hand, changes the initial theory according to the effects of each action and then checks whether the query holds in the updated theory. Incidentally, both come with tradeoffs.

Regression is a form of query rewriting, and can be seen as the sentential solution to pre-image computation for state-based abstractions seen in model checking. This becomes very useful in automated planning, where we may need to check which of a set of action sequences enable the goal. The procedure is general, and works for any basic action theory; formally:[1]

---

[1] Proofs are omitted almost entirely in this book. The bibliographic sections at the end of every chapter detail where formal statements (and their proofs by extension) can be found.

Theorem 3.2 *Suppose $\mathcal{D}$ is any action theory, ϕ any situation-suppressed formula and α any action sequence. Then:*

$$\mathcal{D} \models \phi[do(\alpha, S_0)] \text{ iff } \mathcal{D}_0 \cup \mathcal{D}_{una} \models \phi',$$

where $\phi' = \mathcal{R}[\phi[do(\alpha, S_0)]]$ is the (regressed) formula only mentioning the situation term S_0.

Here, $\mathcal{D}_{una}$ is a set of unique name axioms for actions, that establishes, for example, that *move* is distinct from *reverse*. And, $\mathcal{R}$ is an operator that uses the precondition and successor state axioms to reduce ϕ to a formula ϕ' that says what must have been true initially for ϕ to be true after α. (The precondition axioms are used when the formula to be regressed mentions *Poss*.) We will not give the operator in full, since we will revisit both regression and progression in the presence of probabilities, but rather do a simple example.

Example 3.3 Let us consider the robot example from earlier. We know that the basic action theory, with the initial theory $h(S_0) = 5$, entails:

$$h(do(move, do(move, S_0))) = 3$$

Suppose we slightly modify this example, and we are interested in checking whether $\mathcal{D}$ entails:
$$h(do(move, do(move, S_0))) \neq 4.$$

Regression is defined inductively, with connectives handled in a modular fashion. It eliminates one action in every step:

$$
\begin{align}
& \mathcal{R}[h(do(move, do(move, S_0))) \neq 4] & \text{(i)} \\
=\ & \mathcal{R}[\neg(h(do(move, do(move, S_0))) = 4)] & \text{(ii)} \\
=\ & \neg\mathcal{R}[(h(do(move, do(move, S_0))) = 4)] & \text{(iii)} \\
=\ & \neg\mathcal{R}[h(do(move, S_0)) = 5] & \text{(iv)} \\
=\ & \neg\mathcal{R}[h(S_0) = 6] & \text{(v)} \\
=\ & \neg(h = 6)[S_0].
\end{align}
$$

Here, in step (iv), observe that $h(do(move, S_0)) = 5$ iff $h(do(move, do(move, S_0))) = 4$ by means of the successor state axiom. The same reduction also applies to step (v). Notice also that the successor state axiom for h mentions both *move* and *reverse*, and we using $\mathcal{D}_{una}$ to resolve the appropriate disjunct in the right-hand side (RHS) of that axiom.

With the final step, we are to check if $(h = 5)$ entails $\neg(h = 6)$, and it clearly does. Therefore, by the regression theorem, we have established that: $\mathcal{D} \models h(do(move, do(move, S_0))) \neq 4$.

In sum, $\mathcal{R}$ has a clear and simple specification. The only drawback is that as we consider action sequences of longer length, we should expect that the length of the regression formula grows appropriately. Moreover, if the successor state axiom has *contexts*, which are properties that are to hold in the current situation, they will appear in the regressed formula. Consider, for example, the successor state axiom for breaking an object:

$$Broken(x, do(a, s)) \equiv (a = drop(x) \wedge height(x, s) > 5 \wedge Fragile(x)) \vee (a \neq fix(x)).$$

This says that the object is broken if its dropped at a height of more than 5 feet, and its fragile. Fragile is a rigid property, and so, it does not have a situation argument. It remains broken if its not fixed. Regressing a formula with the *Broken* predicate leads to a formula that will mention the functional fluent *height* and the rigid predicate *Fragile*. While the latter will not have successor state axioms, of course, when regressing the formula further, the presence of $height(x, s)$ might lead to the inclusion of formulas occurring as contexts in the successor state axiom of $height(x, s)$. In general, it is possible that regressing ϕ will result in ϕ' that is exponential in the length of the action sequence. However, it can also be shown that by restricting the syntax of successor state axioms, e.g., only allowing rigids in the context, the regressed formula is only linear in the length of the sequence.

In the general setting, arguably, the exponential blow-up is not a situation we might commonly find ourselves in, at least not for most action theories of interest, even if we are planning a few steps ahead. Nonetheless, we might imagine long-lived agents like robots and virtual assistants. After planning and achieving one of their sub-goals, they may move on to another environment, where other sub-goals become relevant. It is then unnecessary to keep track of the entire history of actions, leading all the way back to point they were turned on. Lin and Reiter [158] argue that the notion of *progression*, that of continually updating the current view of the state of the world, is better suited in these situations. Indeed, many planning and robotics formalisms include a notion of updating their database and beliefs against actions that have already occurred. In the expressive language of the situation calculus, Lin and Reiter prove:

Theorem 3.4 *Suppose $\mathcal{D}$ is any action theory, ϕ any situation-suppressed formula and α any action sequence. Then:*

$$\mathcal{D} \models \phi[do(\alpha, S_0)] \text{ iff } \mathcal{D}'_0 \cup \mathcal{D}_{una} \models \phi[S_0],$$

where $\mathcal{D}'_0$ is the progression of $\mathcal{D}_0$ w.r.t. α.

They show that progression is always second-order definable, and unfortunately, in general, it appears that second-order logic is unavoidable. This is not ideal, as we want to avoid using second-order theorem provers. But Lin and Reiter also identify some first-order definable cases by syntactically restricting situation calculus basic action theories. For example, like with regression, when restricting the contexts of basic action theories to be rigids, progression becomes first-order definable.

Example 3.5 To see progression in action with the robot example, suppose, we are again interested in whether:

$$\mathcal{D} \models h(do(move, do(move, S_0))) \neq 4.$$

We omit the details here, but it can be shown that the progression of $(h = 5)[S_0]$ w.r.t. the action sequence $[move \cdot move]$ is $(h = 3)[S_0]$. Since the robot has moved forward by 2 units, its distance to the wall is reduced by 2 units, not surprisingly. By the progression theorem the above entailment can be checked instead with:

$$(h = 3)[S_0] \cup \mathcal{D}_{una} \models (h \neq 4)[S_0].$$

This is clearly true, and so the projection query is indeed entailed.

3.2.6 A Programming Language

With a domain axiomatization in hand, the most classical and well-studied application is *automated planning*. Given a goal ϕ, find a sequence of actions α such that it both *executable* and goal enabling:

$$\mathcal{D} \models Executable(\alpha, S_0) \wedge \phi[do(\alpha, S_0)].$$

Here, the executability property is an abbreviation for:

$$\bigwedge_{i=1}^{n} Poss(a_i, do([a_1, \ldots, a_{i-1}], S_0)).$$

That is, $Poss(a_1, S_0)$ holds, $Poss(a_2, do(a_1, S_0))$ holds, and so on.

One obvious way to find the desired sequence is try to check for $n = 1, 2, \ldots$, whether:

$$\mathcal{D} \models \forall a_1, \ldots, a_n. \ Executable([a_1, \ldots, a_n], S_0) \wedge \phi(do([a_1, \ldots, a_n], S_0)).$$

We would stop the check for the smallest n such that executability and goal satisfaction is true.

Note that checking the executability as well as whether ϕ is true after α are regressable. But it still requires a blind search over all possible action sequences of length n. The routine can be improved by a number of strategies. For example, we could avoid considering actions that do not affect any of the predicates mentioned in the goal; a notion of *relevancy* could be established. Moreover, we may want to restrict the expressiveness of the initial theory to possibly avoid a general first-order theorem prover. While the effectiveness question is interesting to consider, the overall regime is nonetheless limited in having the agent function purely as an entity that satisfies a goal. We might want to instead guide the agent's operation by interleaving specified behavior and planning, where further, the specified behavior could be a partial description.

This motivated the development of GOLOG, a high-level programming framework based on the situation calculus. Like standard programming languages, there are if-then-else and while constructs, but the key distinction is that the conditions of these constructs are situation calculus formulas. Atomic programs are variable assignments in standard programming languages, but in our setting, these are actions that the agent can perform. Formally, GOLOG programs are defined over the following constructs:

$$a \mid \phi? \mid (\delta_1; \delta_2) \mid (\delta_1 \mid \delta_2) \mid (\pi \ x)\delta(x) \mid \delta^*$$

standing for atomic actions, tests, sequence, nondeterministic branch, nondeterministic choice of argument and nondeterministic iteration respectively. Other constructs can then be defined in terms of these:

$$\textbf{if } \phi \textbf{ then } \delta_1 \textbf{ else } \delta_2 \textbf{ endIf} \doteq [\phi?; \delta_1] \mid [\neg\phi?; \delta_2]$$

$$\textbf{until } \phi \textbf{ do } \delta \textbf{ endUntil} \doteq [\neg\phi?; \delta]^*; \phi?$$

$$\textbf{while } \phi \textbf{ do } \delta \textbf{ endWhile} \doteq \textbf{until } \neg\phi \textbf{ do } \delta \textbf{ endUntil}$$

A semantics is given to such programs using an operator $Do(\delta, s, s')$, which is to be understood as saying δ starts in s and *terminates* in s'. For example, for an atomic action:

$$Do(a, s, s') \doteq Poss(a, s) \wedge s' = do(a, s).$$

That is, a is executable in s, and doing a in s leads to the successor s'. We omit the details of the full semantics, which involves second-order constructs to interpret nondeterministic iteration (and consequently, while loops), but note the following contrast to planning. What we viewed as a straightforward approach to planning could be reformulated using:

$$\textbf{until } \phi \textbf{ do } \pi a. \ a \textbf{ endUntil,}$$

where until the goal is made true, the agent picks an action a, and performs it, leading to an executable sequence of actions enabling the goal. A more interesting alterative to avoid the blind search by providing expert knowledge before, after and during the search, say:

$$a_1|a_2; [\textbf{until}\neg\phi \textbf{ do } \pi a. \ Acceptable(a)?; a \textbf{ endUntil}]; a_3.$$

That is, the modeler thinks the agent may want to nondeterministically choose between a_1 or a_2 before search, and a_3 after the search. The search itself is decorated with additional constraints via the predicate *Acceptable* that determines which actions should be considered at all for the search. For example, in the simplest case, if we are attempting to get the robot to the wall, we might get the robot to avoid any action not related to moving forwards or backwards. More generally, procedural knowledge can be complex and powerful. Moreover, we might associate costs or rewards with actions, in the sense that we need the robot to achieve goals with the lowest costs or the highest rewards. We could then further empower program execution to respect such quantities and get the robot to behave *optimally*.

3.3   Technical Devices

The book is about a working proposal for logic, probability and actions entirely in first-order logic. However, now that the prelude to the classical situation calculus is over, we will be modelling knowledge, belief and noise. For this, we will introduce four technical devices, all of which are also formalized in first-order logic.

For dealing with knowledge and degrees of belief, in particular, we will need to consider not just one possible world but many, as the agent may be uncertain about the state of the world. This will require us to consider initial situations other than S_0. Second, since degrees of belief will involve manipulating probabilities, we will require first-order models to obey standard arithmetic symbols and quantities such as $+$, $\times$ and Euler's number e. Third, we define a logical term standing for summation in the usual mathematical sense, that is to be understood as an abbreviation for a formula involving second-order quantification. This will become useful for discrete probability distributions. And fourth, we analogously define a logical term standing for integration in the usual mathematical sense, which will be used for continuous probability distributions.

3.3.1   Many Initial Situations

As noted, the constant S_0 is assumed to give the actual initial state of the domain, but with knowledge and belief, the agent may consider others possible. The presence of multiple possible world capture the beliefs but also the ignorance of the agent, which we will get into in the next chapter. For now, however, it suffices to introduce a predicate *Init* taking a situation term with the understanding that it is a situation without a *predecessor*:

$$Init(s) \doteq \neg \exists a, s'. s = do(a, s').$$

The picture that emerges is that we have a set of tree-like structures, each with an initial sitation at the root, and whose whose edges are actions. These initial situations are likely to disagree not only on what is true initially, but also on what is true after actions. That is, you may have two initial situations that agree on the truth of all formulas, but differs on what is true after doing an action. In terms of convention, we use ι to range over such initial situations only.

Note that the foundational axioms would need to be adjusted to allow for multiple initial situations. This is accomplished using the following induction axiom:

$$\forall P, s. \; [Init(s) \supset P(s)] \wedge \forall a, s. \; [P(s) \supset P(do(a, s))] \supset \forall s. \; P(s).$$

That is, we simply scoped the base case to include all initial situations rather than just S_0 in the version from Sect. 3.2.3.

3.3.2 $\mathbb{R}$-Interpretations

For our purposes, the notion of entailment will be assumed w.r.t. a class of Tarskian (first-order) structures that we call $\mathbb{R}$-interpretations.

Definition 3.6 By $\mathbb{R}$-interpretation we mean any $\mathcal{L}$-structure where $\{=, 0, 1, +, \times, /, -, e, \pi, <, >\}$, exponentiation and logarithms have their usual interpretations.

That is, "$1 + 0 = 1$" is true in all $\mathbb{R}$-interpretations, if "$x > y$" is true then "$\neg(y > x)$" is true, and so on. So, henceforth, when we write $\mathcal{D} \models \phi$, we mean that in all $\mathbb{R}$-interpretations where $\mathcal{D}$ is true, so is ϕ. Note, for example, that natural numbers can be defined in terms of a predicate by appealing to $\mathbb{R}$-interpretations. Let

$$Natural(x) \doteq \forall P[(P(0) \wedge \forall x(P(x) \supset P(x + 1))) \supset P(x)].$$

Theorem 3.7 *Let M be any $\mathbb{R}$-interpretation, and c a constant symbol of $\mathcal{L}$. Then, $M \models Natural(c)$ iff $c^M \in \mathbb{N}$.*

Recall that for any $\mathcal{L}$-term t and $\mathcal{L}$-interpretation M, we use t^M to mean the domain element that t references.

What $\mathbb{R}$-interpretations allows us is to talk about numbers in $\mathcal{L}$-sentences and understand operations over them in the usual mathematical sense.

3.3.3 Summation

When reasoning about degrees of belief with discrete probability distributions, we will need to capture finite summations. Here we show how such summations can be characterized as an abbreviation for an $\mathcal{L}$-term using second-order quantification. Of course, we do not intend on using second-order theorem provers to reason about these sums. Rather we simply wish to show that they are not meta-language constructs, and can be semantically studied just like any other formal object in the language.

Let f be any $\mathcal{L}$-function from $\mathbb{N}$ to $\mathbb{R}$. Let SUM(f, n), standing for the sum of the values of f for the argument 1 through n, be defined as an abbreviation:

$$\text{SUM}(f, n) = z \doteq \exists g[g(1) = f(1) \wedge$$
$$g(n) = z \wedge$$
$$\forall i\, (\, 1 \leq i < n \supset g(i + 1) = g(i) + f(i + 1)\,)].$$

The variable i is understood to be chosen here not to conflict with any of the variables in n and z. The function g from $\mathbb{N}$ to $\mathbb{R}$ is assumed to not conflict with f. (That is, the logical terms are distinct.)

This can then be argued to correspond to summations in the usual mathematical sense as follows:

Theorem 3.8 *Let f be a function symbol of $\mathcal{L}$ from $\mathbb{N}$ to $\mathbb{R}$, c be a term, and n be a constant symbol of $\mathcal{L}$. Let M be any $\mathbb{R}$-interpretation. Then,*

$$if \sum_{i=1}^{n^M} f^M(i) = c^M \ then \ M \models \text{SUM}(f, n) = c.$$

Henceforth, we write:

$$\sum_{i=1}^{n} t$$

to mean the logical formula SUM(t, n) for a logical term t. Here, i is assumed to not conflict with any of the variables in n and t.

It is worth noting that this logical formula can be applied to summation expressions where the arguments to the terms are not restricted to natural numbers, but taken from any finite set. For example, suppose H is any finite set of terms $\{h_1, \ldots, h_n\}$. We can then use terms such as:

$$\sum_{h \in H} t(h)$$

standing for an abbreviation, similar to SUM(t, n): let g be a function, and let $g(i) = t(h_i)$. Then clearly the above sum defines the same number as $\sum_{i=1}^{n} g$. In the sequel, we sometimes sum over a finite set of situations, or a finite vector of values, which is then understood as an abbreviation in this sense.

3.3.4 Integration

Finally, we characterize integrals as logical terms. We begin by introducing a notation for limits to positive infinity. For any logical term t and variable x, we introduce a term characterized as follows:

$$\mathrm{LIM}[x, t] = z \;\doteq\; \forall u (u > 0 \supset \exists m \, \forall n (n > m \supset |z - t_n^x| < u)).$$

The variables u, m, and n are understood to be chosen here not to conflict with any of the variables in x, t, and z. The abbreviation can be argued to correspond to the limit of a function at infinity in the usual sense:

Lemma 3.9 *Let g be a function symbol of $\mathcal{L}$ standing for a function from $\mathbb{R}$ to $\mathbb{R}$, and let c be a constant symbol of $\mathcal{L}$. Let M be any $\mathbb{R}$-interpretation of $\mathcal{L}$. Then we have the following:*

$$\text{If } \lim_{x \to \infty} g^M(x) = c^M \quad \text{then } M \models (c = \mathrm{LIM}[x, g]).$$

Henceforth, we write:

$$\lim_{x \to \infty} t$$

to mean the logical formula $\mathrm{LIM}[x, t]$.

Next, for any variable x and terms a, b, and t, we introduce a term $\mathrm{INT}[x, a, b, t]$ denoting the definite integral of t over x from a to b:

$$\mathrm{INT}[x, a, b, t] \;\doteq\; \lim_{n \to \infty} h \cdot \sum_{i=1}^{n} t_{(a + h \cdot i)}^{x}$$

where h stands for $(b - a)/n$. The variables are chosen not to conflict with any of the other variables.

Lemma 3.10 *Let g be a function symbol of $\mathcal{L}$ standing for a function from $\mathbb{R}$ to $\mathbb{R}$, and let a, b, c be constant symbols of $\mathcal{L}$. Let M be any $\mathbb{R}$-interpretation of $\mathcal{L}$. Then we have the following:*

$$\text{If } \int_{a^M}^{b^M} g^M(x)\, dx = c^M \quad \text{then } M \models (c = \mathrm{INT}[x, a, b, g]).$$

Finally, we define the definite integral of t over all real values of x by the following:

$$\int_x t \;\doteq\; \lim_{u \to \infty} \lim_{v \to \infty} \mathrm{INT}[x, -u, v, t].$$

The main result for this logical abbreviation is the following:

Theorem 3.11 *Let g be a function symbol of $\mathcal{L}$ standing for a function from $\mathbb{R}$ to $\mathbb{R}$, and let c be a constant symbol of $\mathcal{L}$. Let M be any $\mathbb{R}$-interpretation of $\mathcal{L}$. Then we have the following:*

$$\text{If } \int_{-\infty}^{\infty} g^M(x)\, dx = c^M \quad \text{then } M \models \left(c = \int_x g(x) \right).$$

The characterization of integrals for a many-variable function f, from $\mathbb{R}^k$ to $\mathbb{R}$, is then an easy exercise, which we omit here.

3.4   Notes

The underlying language of the situation calculus has received a lot of attention in the action community. Reiter's book [190] is perhaps the best comprehensive coverage on various extensions to the situation calculus, including for time and decision theory. But, of course, there has been considerable development since then, such as abstracting action theories [10], reconstructing the language in a modal logic [136, 226] (which we will revisit in a penultimate chapter), as a planning formalisms for generating recursive plans [117, 156], reasoning about multiple agents [122], and so on. Reiter's book also systematically introduces the frame problem, foundational axioms, the projection problem, and solutions to that problem: regression and progression.[2] The second-order nature of progression was further studied in [229].[3]

For Drew McDermott's quote about Reiter's reformulation, see [168].[4] For different approaches to modelling time in the situation calculus, such as introducing a new sort versus linking to action execution, see [134, 182, 190].

The programming language GOLOG was introduced in [146]. For a vision of cognitive robotics built on this model, see the discussion in [138]. Much of our discussion on planning and programs in this chapter is adapted from that article. The modelling approach here is via *axioms,* that is, sentences in the language capturing the domain of interest.[5] For the use of GOLOG to augment plan search, see [9]. For an account of rewards and optimal behavior in GOLOG, see [34].

There are, of course, alternate formalisms, such as the fluent calculus [214] and other closely related approaches, such as those based on dynamic logics [64, 65, 88, 226]. Planning languages are special-purpose languages designed primarily to describe actions, effects and goals (and/or rewards), and synthesize plans effectively with those descriptions [85, 118]. Thus, they are related to but significantly less expressive than logics for action, which admit reasoning about and comparing past and future events, often with histories being first-class citizens. Planning languages are also deliberately propositional for computational reasons. See also [37, 190], among others, for the considerations and tradeoffs in planning languages

---

[2] Regression is a form of query rewriting [231], and can be seen as the sentential solution to pre-image computation for state-based abstractions seen in model checking [46].

[3] Since their work, a number of other special cases have been studied [160].

[4] There are numerous nonmonotonic accounts that rely on a controlled extension to predicates—often referred to as *model minimization*—to overcome the need to specify every qualifier and condition, e.g., [3, 87].

[5] This can be contrasted to a model checking approach, where the application of interest is captured directly as a propositional, first-order or modal structure [107, 227].

when contrasted to first-order knowledge representation languages. For some investigations on the use of first-order logic and its probabilistic extensions for capturing common sense reasoning, see [55, 151, 197].

Finally, in this chapter, given a non-negative real-valued function, our notion of an integral of this function is based on the Riemann integral [222], in which case the function is said to be *integrable*. There are limitations to the Riemann integral; for example, the function $f : [0, 1] \rightarrow \mathbb{R}$ where

$$f(x) = \begin{cases} 1 & \text{if } x \text{ is rational} \\ 0 & \text{otherwise} \end{cases}$$

is not integrable in the Riemann account. In the calculus community, generalizations to the Riemann integral, such as the gauge integral [211], have been studied that allow for the integration of such functions. We have chosen to remain within the framework of classical integration, but other accounts may be useful.

Knowledge

4

*Either we shall find what it is we are seeking or at least we shall free
ourselves from the persuasion that we know what we do not know.*

—**Plato**, *The Republic*

The situation calculus is a formalism for reasoning about actions and effects, and it is implicit in the axiomatizations that these effects are actualized in the world. It is also implicit in such axiomatizations that the initial theory represents the things we know to be true, and so the actions are assumed to be changing these things. But who is the *knower*?

In a database setting, the initial theory captures the facts that are entered in the database, and so the actions are updates performed on the database itself. It does seem reasonable to think, that the facts in the database are in agreement with reality. In that sense, the database implicitly *knows* the world (or at least the part of the world that the database is capturing).

This simple notion breaks down in most applications involving artificial agents. Except in controlled environments like video games where non-player characters would have a full picture of the virtual world, uncertainty is a fact of life. Robots can be expected to have only a partial view of the world, limited by what was programmed initially and what they observe. If there are multiple agents in the environment, a robot might not only be uncertain about the physical world, but also about what the other agents are capable of, and what information they possess. In fact, even with databases, if we allow that the information entered in the database might be erroneous or noisy, we might want to entertain a notion of informational incompleteness, or *not-knowing*.

© The Author(s), under exclusive license to Springer Nature Switzerland AG 2023 49
V. Belle, *Toward Robots That Reason: Logic, Probability & Causal Laws*,
Synthesis Lectures on Artificial Intelligence and Machine Learning,
https://doi.org/10.1007/978-3-031-21003-7_4

4.1   Truth and Knowledge

Now that we opted to expand the set of modeling concepts to include *knowledge*, it is worth taking an informal tour of the arena and its players.

4.1.1   Objective and Subjective Sentences

There are things that are true, and then there is what is known. The former is *objective*, but the latter is a *subjective* view of what is true. Martian soil contains chlorine, but this may or may not be known by the robot at hand. The robot's mental model might not even include planetary designs and so the term 'Mars' itself might not be known.

Thus, logically, we will distinguish between atoms that are true, versus what is known by the robot. And, from an epistemological viewpoint it is possible for these atoms to be mutually inconsistent, although we will not delve into that further. (For example, it is possible that the robot might know false things about Mars, because, say, the engineer included outdated information about the red planet in the robot's database.)

For our part, we will assume that the robot simply does not know things about the world, and so over the course of its operation, its ignorance is reducing where possible. To its benefit, if the robot knows a fact, then that will also be true in the actual world. So the robot might not know if there is chlorine in Martian soil, but know that the equatorial radius of Mars is greater than 3000 km, which happens to be true in reality.

All this leads to a notion of *knowledge expansion*, where the robot is gradually knowing more as it operates, together with the notion that *knowledge is truth*.

4.1.2   Actions

The next issue at hand is to understand the relationship between objective sentences and subjective sentences, on the one hand, and *actions*, on the other.

Until now, we have assumed actions to affect changes in the world, so they are *physical* in nature, even if the world itself is virtual. A database entry being updated, a non-player character in a video game attacking the human player, and so on, are all instances of actions being executed that change something about the world. But how do such actions affect the agent's knowledge?

To answer that question let us consider a simple but abstract picture, seen in Fig. 4.1. In so much as we could depict the world as a set of facts, incomplete knowledge can be realized as a set of worlds. Each world differs from another in at least one fact, so no two worlds are exactly the same. Let us suppose Fig. 4.1 is saying that although the world is really red, the robot is not sure, and thinks it might be red, brown or blue.

Fig. 4.1 The actual world versus a set of possible worlds entertained by the robot

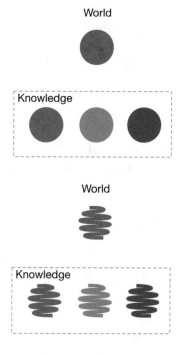

Fig. 4.2 An action that blurs the coloring

A physical action changes the world, so suppose there is an action that blurs the coloring. The actual world is affected, of course, but as well, every world considered possible by the robot is affected too. This yields Fig. 4.2.

But why would the worlds considered possible by the robot be affected at all, or affected in the exact same way? For one thing, we are assuming the action was observable to the robot (and presumably performed by the robot itself). Of course, in other situations, we might want to deal with actions performed by other agents or even *nature*, all of which are outside the control of the agent. So, these may not be observed. Thus, even though the world changes the agent may not be aware actions have occurred, and the possible worlds might be outdated w.r.t. reality. Thus, we would need a framework for exogenous actions, but will not delve into that here.

So suppose the agent is aware of the action taking place, and executed the action itself. Its worlds would change in a manner entirely consistent with the actual world *only if* all these worlds obey the same dynamic laws, a notion we will make precise shortly. Indeed, if dropping a ball in one the possible worlds makes the ball go up because gravity works strangely in that world, then the action would not affect the world in the same way as the actual world.

To sum up, we will make three assumptions for simplicity: (a) all actions are performed by the robot, (b) all worlds obey the same dynamic laws, and (c) actions will always succeed and affect the intended change. To drop (a), we would need an account of exogenous actions or multiple agents, which is very doable, but we will not consider that for the sake of simplicity. There is no inherent technical limitation that insists on (b), but it will complicate the treatment. So again for the sake of simplicity, the matter is ignored. As for (c), this is considered in the next chapter.

4.1.3 Sensing

If the only kind of actions available update worlds, the number of possible worlds would never change. The robot's knowledge updates w.r.t. those changes, in the sense of considering the colors of all worlds to be blurred, but would never come to know which of the three is the actual world.[1] Here is where *sensing actions* come to the rescue. These are distinct from physical actions is not affecting any *physical* change in the worlds, but rather reveal some property about the actual world.

Consider a sensing action that identifies the color of the world. On executing the action, the robot immediately infers that since the actual world is red, the brown and blue worlds cannot be possible. Starting from Fig. 4.2, such an action yields Fig. 4.3, where two of the worlds are discarded.

It is conceivable to imagine actions that conflate the two kinds of actions, that is, actions that affect the world but also reveal information about the world. For example, when smashing a bottle of wine of the table, we break the bottle and spill the wine on the table but this might also reveal that the wine is white and not red. The latter effect is solely affecting the mental state and does not change the actual world. But we will keep physical and sensing actions distinct for the sake of simplicity.

To wrap up this informal exposition, let us recap that there are objective knowledge and subjective knowledge, and while physical actions affect the truth values of both, sensing actions only affects the subjective one. Moreover, differently from the classical situation calculus, it is now possible to enhance the axiomatization as well as the programming language with epistemic constructs. The precondition for opening a safe is *knowing* the number code for the safe, for example. This allows for *meta-level* control to be conditioned on what the robot knows at the time of acting.

---

[1] A minor remark: physical actions might affect these worlds in a manner that leaves some worlds semantically identical to others; the agent knows more as a result, and we could view this as a case of reducing the number of possible worlds. For example, imagine the robot colors the world red. Then regardless of the initial color, the robot does not need observations to know that the world is red.

Fig. 4.3 Sensing the color of
the actual world

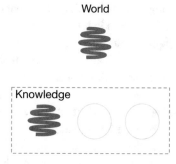

4.2   The Knowledge Macro

The informal picture discussed leads itself to a nice formal apparatus, that of the *possible worlds semantics* for knowledge.

4.2.1   Possible Worlds

Following the works of Kripke, Hintikka and others, we can augment propositional logic with a *modality* for knowledge. Suppose w is a propositional interpretation, or a *world*: that is, it assigns truth values to the propositional variables in the language. Let $\mathcal{W}$ be the set of all worlds, that is, it is the set of all propositional interpretations for those variables. Let $\mathcal{K}$ be a binary *accessibility* relation on $\mathcal{W}$. By means of this relation, pairs of worlds such as (w, w'), (w', w'') and (w, w'') capture the robot's epistemic possibilities. For example, $\mathcal{K}(w, w')$ says that the robot cannot distinguish between these two interpretations. Put differently, if the robot is actually operating in w, it considers the possibility that it is in w'. This is a reflection of the informational incompleteness of the robot's mental state.

Using $\models$ as usual for satisfaction, we can interpret an extended language that includes propositional formulas ϕ together with formulas of the sort $\boldsymbol{K}\phi$, $\neg\boldsymbol{K}\phi$ and $\boldsymbol{K}\phi \wedge \neg\boldsymbol{K}\neg(\phi' \vee \phi'')$. Here, $\boldsymbol{K}$ is a *modal operator*, that is, the epistemic modality.

What is the relationship between $\boldsymbol{K}$ and $\mathcal{K}$? The pair $(\mathcal{K}, w)$ is an interpretation for the extended language, just as w interpreted the propositional language. We would say:

- $(\mathcal{K}, w) \models \boldsymbol{K}\phi$ if and only if for every w' such that $\mathcal{K}(w, w')$, we have $(\mathcal{K}, w') \models \phi$.

So, when treating w as the actual world, for ϕ to be *known*, it must be the case that ϕ is true at all worlds considered possible at w. We read $\boldsymbol{K}\phi$ as *knows ϕ*.

It is worth briefly recalling that the knower in question is the robot, and not the human designer or engineer. When looking at reasoning problems, for example, the human designer might be in a position to provide objective information, and so the epistemic modality is always used in relation to that, in terms of what the robot knows in a given situation. This will be illustrated shortly.

We previously suggested that we will want to assume knowledge to be true. Which ingredient of the construction allows for this? In the general way $\mathcal{K}$ was set up above, there is none. However, suppose we insist that w itself should be possible from w, that is, $\mathcal{K}(w, w)$. Then for any $\boldsymbol{K}\phi$ that is true at $(\mathcal{K}, w)$, it must be the case that ϕ is also true at w. This is because by definition, ϕ must come out true at all worlds w' such that $\mathcal{K}(w, w')$, which includes w. By extension, it follows that if $\mathcal{K}$ is *reflexive*, $\boldsymbol{K}\phi \supset \phi$ is a valid formula.

Conversely, we would say that $\boldsymbol{K}$ could be interpreted as *belief* in the absence of the reflexivity, to indicate a weakness in contrast to knowledge. An agent could believe ϕ but then there would no requirement that ϕ actually hold in the actual world. For the sake of simplicity, we will assume reflexivity but use the terms *knowledge* and *belief* interchangeably.

What other properties might be useful for knowledge to have? Two very popular properties are *positive* and *negative introspection*. That is, knowing ϕ means the robot knows that it knows ϕ: consider this a meta-level appreciation. Analogously, not knowing ϕ means that the robot knows that it does not know ϕ.

Like with knowledge being true, introspection reflects constraints on $\mathcal{K}$. Positive introspection equates to transitivity, which means for all w, w', w'', if $\mathcal{K}(w, w')$ and $\mathcal{K}(w', w'')$ then $\mathcal{K}(w, w'')$. Negative introspection equates to the Euclidean property, which means for all w, w', w'', if $\mathcal{K}(w, w')$ and $\mathcal{K}(w, w'')$ then $\mathcal{K}(w', w'')$.

4.2.2 The Ideal Reasoner

Obviously, such properties imply a powerful, and ideal, reasoner. Negative introspection, in particular, implies meta-level knowledge of ignorance, which means the robot can immediately try to close gaps in its knowledge, if it so wishes and there are available means to do that. In an intelligent and fully autonomous robot, this can come across as alarming but also quixotic.

The other major concern that could be raised against the power of the reasoner is the fact that the robots are *logically omniscient*. Suppose $\phi \supset \phi'$ is valid in propositional logic. By construction, if $\boldsymbol{K}\phi$, then $\boldsymbol{K}\phi'$ as well, so the robot knows all the consequences of its knowledge. In particular, if there is an involved statement that is valid in a logic (such as Fermat's last theorem or similar), then in the modal extension, the robot would know that statement. This is because the statement must be true in all the worlds considered possible owing to its validity, and therefore, the robot knows it. This might also come across as alarming if it was at all feasible.

The feasibility issue is this: if we wanted to show that in a logic that, for example, includes the Peano axioms, we can prove $\boldsymbol{K}\psi$, where ψ is some involved consequence of those axioms, we would need a theorem prover. So even if it were the case that $\boldsymbol{K}\psi$, inferring that in practice could take exponential time. Thus, although a variety of statements involving introspection and complicated mathematical theorems can indeed be expressed in the logic, using such sentences either as part of the premise or the conclusion for the robot's reasoning would not likely to lead to *real-time* acting with modest computing resources.

If its any comfort, much of this discussion carries over from the non-epistemic setting in the previous chapter. Indeed, complicated mathematical theorems can be expressed in predicate logic, and using such sentences either as part of the premise or the conclusion for the robot's reasoning would not be sensible. From the general specification language, either the implementation language needs to be controlled, by insisting that the axiomatization only use, say, Horn logic. Otherwise, the axiomatization and programming assumes designer discretion, to ensure that conditions in a program or projection problems do not involve contrived, Sudoku-puzzle-type expressions. Unless the situation demands it, of course: the robot might actually need to solve a puzzle, for example, to achieve its goals.

The modal setting can make the reasoning harder: for a propositional language, the satisfiability problem with the epistemic operator where $\mathcal{K}$ is reflexive, transitive and Euclidean is also NP-complete, as in standard propositional logic. However, dropping negative introspection, for example, makes the satisfiability problem PSPACE-hard. A multi-agent environment can be modeled by allowing every agent to have its own accessibility relation $\mathcal{K}$, and this also affects the complexity of reasoning. The case of a reflexive, transitive and Euclidean $\mathcal{K}$ but with two or more agents means the satisfiability problem is PSPACE-complete. A further extension to the multi-agent setting is the notion of *common knowledge*: we say ϕ is common knowledge when each agent knows ϕ, each agent knows that every agent knows ϕ, and so on, ad infinitum. Common knowledge is the epistemic operator needed to model the muddy children puzzle and the Byzantine general's communication problem. The satisfiability problem with common knowledge and two or more agents is EXPTIME-complete.

In this book, we will limit ourselves to the single agent case, but clearly one of the benefits of having an epistemic operator in the language is to reason with multiple agents. The formal language allows such extensions to be explored in a natural, but simultaneously general, way.

Lastly, let us note that we have discussed the modal semantics at a propositional level for initiatory purposes. The situation calculus, on the one hand, will allow us to reason about dynamics, to capture the act-sense we motivated using the red, brown and blue worlds. And on the other, we will be able to reason about knowledge in a quantificational setting. This will mean the capability to express two types of interesting statements. First, knowledge about quantified sentences: the robot knows that all humans are mammals. Second, quantifying in: there is a person that the robot knows is both short and a professional basketball player. Quantifying-in is intriguing because the statement above implies that there is at least one witness to "short and a professional basketball player", and so the robot knows the identity of the person. This is contrasted with: the robot knows that there is a person who is both short and a professional basketball player. In this latter phrasing, the quantified sentence is interpreted solely at the possible worlds, so the witness may differ from world to world. What this means is that it may not be possible for the robot to identify who this individual is, but simply know that such an individual exists.

One remark before the definitions: because we will be exploring the first-order epistemic and dynamical setting in the same formal framework as from the previous chapter, it will mean that well-formed sentences about knowledge will be a bit different from classical epistemic logic. In "standard" epistemic logic, the worlds are not actually part of the language, and indeed we write sentences such as $K\phi$ and $K\phi \supset \neg KK\phi'$ without making a reference to the actual world or the other worlds considered possible. But in the classical situation calculus, situations represented the changing world, all identified w.r.t. a distinct initial situation for the empty history S_0. In the epistemic setting, we will therefore think of multiple initial situations, and so these situation terms will be represented in the language. Logicians would say the worlds are *reified* in the language.

This linguistic idiosyncrasy does not affect the kinds of things we are able to reason about. It just requires some getting used to, especially for readers already familiar with standard epistemic logic. In a later chapter, we will revisit this issue, and recast the situation calculus in a manner that closely resembles the standard variant.

4.2.3 The Epistemic Fluent

The possible-worlds semantics for knowledge is based on the notion that there many different ways the world can be, where each world stands for a complete state of affairs. In the situation calculus, the idea is that situations can be viewed as possible worlds. This insight is due to Moore, which was later refined by Scherl and Levesque.

But situations are histories of actions, so how do we define knowledge initially, and further, how do we define knowledge after actions and sensing? The key intuition is that we define a possible state of affairs *initially*, that is, before any of the actions have been performed. This allows us to talk about what is known by the robot at the start. By means of the axiomatization, after an action, we will repurpose the accessibility relation for all of the successor situations considered initially. So, after an action, what is known is defined in terms of these successor situations. This leads to the robot not only knowing what it did at the start, but also knowing the updated set of facts affected by the action that just performed.

To see how this all works out, let us leave aside the initial worlds as well as actions for the moment, and simply consider how the formalism captures knowledge. The situation calculus was introduced as a dialect of first-order logic, with three sorts: situations, actions and objects. The domain modeling was achieved using rigids and fluents. The epistemic situation calculus maintains the simplicity of the ontology: we will use a special binary fluent K, taking two situation arguments, for the accessibility relation between worlds. That is, $K(s', s)$ says that when the agent is at s, he considers s' possible. We could see K as the fluent version of the meta-linguistic binary accessibility relation $\mathcal{K}$. Moreover, knowledge is purely a *macro*, in the sense of expanding into a well-formed formula mentioning only K, and not a modal operator per se, standing for truth at accessible worlds:

Definition 4.1 Let ϕ be any situation-suppressed formula. The agent knowing ϕ at situation s, written $Knows(\phi, s)$, is the following abbreviation:

$$Knows(\phi, s) \doteq \forall s'. \, K(s', s) \supset \phi[s'].$$

This definition does not make any reference to the action history, and is based purely on the accessibility relation, much like the standard exposition. The axiomatization, as we will see, is basically about getting K to align with the history. Analogously, properties such reflexivity and transitivity stipulated for $\mathcal{K}$ will be axiomatized for K.

Let us now get comfortable with this macro. To know that John is enrolled in the situation s, we write:

$$Knows(Enrolled(john), s) \doteq \forall s'. \, K(s', s) \supset Enrolled(john, s').$$

The macro also works with open formulas:

$$Knows(Enrolled(x), s) \doteq \forall s'. \, K(s', s) \supset Enrolled(x, s').$$

The macro can also be applied to formulas involving terms:

$$Knows(code(y) = age(mother(x)), s) \doteq \forall s'. \, K(s', s) \supset [code(y, s')$$
$$= age(mother(x), s')].$$

The robot knows that the code of safe y is the same number as the age of the mother of x. Here, notice that while both code and age are fluents, as they can change from situation to situation, a person being the mother of x is situation-independent. Thus, it is a rigid function, and this is reflected in the expansion of the *Knows* macro.

Consider now the notion of quantifying-in discussed earlier. The following sentence:

$$\exists x \, Knows(Close(x) \wedge \exists y \, (y \neq x \wedge age(y) = age(x)), s)$$

says that there is someone the robot knows is close, and moreover, the robot knows that there is another person whose age is the same as the one who is close. The sentence expands to:

$$\exists x \, \forall s'. \, K(s', s) \supset [Close(x, s') \wedge \exists y \, (y \neq x \wedge age(y, s') = age(x, s'))].$$

In other words, the identity of the individual x is known (and is the same) in every world considered possible, whereas the identity of y may differ between the possible worlds, as the witness to that existential is interpreted w.r.t. the situation s'.

4.2.4 Effects of Actions

To specify the behavior of K, and consequently *Knows*, at non-initial situations, a successor state axiom is needed for K (as with any other fluent). However, this one will be domain-independent. This axiom, intuitively, tests whether situations are to remain accessible as actions occur. To understand how that axiom needs to be constructed, let us informally motivate knowledge initially, and the effect of a *physical action* using a single functional fluent f.

Suppose the robot knows that f can take only one or two values initially:

$$Knows(f = 0 \vee f = 1, S_0).$$

As discussed, this really means that at every situation considered accessible from S_0, the value of f is either 0 or 1:

$$\forall s.\ K(s, S_0) \supset f(s) = 0 \vee f(s) = 1.$$

This may correspond to Fig. 4.4. Note that, for simplicity, there are finitely many worlds and finitely many values for f: given the expressiveness of the situation calculus, neither of this need be true.

Suppose we had an action that increments the value of f. This could be enabled by means of the following successor state axiom for f:

$$f(do(a, s)) = u \equiv [\neg \exists z(a = incr(z)) \wedge u = f(s)] \vee [\exists z(a = incr(z) \wedge u = (f(s) + z))].$$

Equivalently, using the IF-THEN-ELSE notation, we have:

$$f(do(a, s)) = \text{IF} \exists z(a = incr(z)) \text{THEN} (f(s) + z) \text{ELSE} f(s).$$

The action $incr(z)$ increments the value of f by precisely z, and so f can be decremented by providing a negative argument to the action. Moreover, $incr(z)$ is the only action affecting the fluent, and so all other actions do not change f's value. This incorporates the monotonic solution to the frame problem.

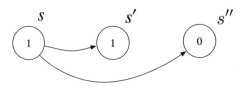

Fig. 4.4 The initial worlds w.r.t. the functional fluent f. The number inside the circles denote the values for f at the situation. Letting $s = S_0$, situations s' and s'' are accessible from s

Fig. 4.5 The action $incr(1)$ is performed. From situations s, s' and s'', we obtain successor situations $do(incr(1), s), do(incr(1), s')$ and $do(incr(1), s'')$

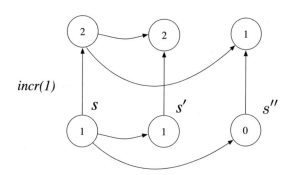

Clearly, performing the action at any given situation increments the value of f. This means that in every situation considered possible initially, the successor situations have f's values appropriately incremented. Then, how does this affect K? Intuitively, knowing that the value of f has incremented would simply mean that if the robot considered a world possible where f took the value n, it will now consider the value of f in that world is $n + z$. Thus, we do not need to tinker with K in any manner. In particular, the action $incr(1)$ yields Fig. 4.5. We could also say what is known formally:

$$Knows(f = 1 \vee f = 2, do(incr(1), S_0)).$$

Let us now consider the effect of sensing actions. Assume the availability of a sensor that checks whether f is zero, by means of the action $sensezero$. To model such capabilities, we need to consider a new type of axioms in the basic action theory called *sensing axioms*, operationalized using a distinguished fluent SF. For $sensezero$, we would have:

$$SF(sensezero, s) \equiv f(s) = 0.$$

The truth value of the fluent at a situation on taking $sensezero$ as an argument is true if and only if f takes the value 0 in that situation.

The intuition is this. Some things are true in the actual world, and the robot will use its sensors to observe those truths. Consequently, some of the worlds considered previously possible need to be discarded if they disagreed with the outcome of the sensor. This functionality is operationalized in the successor state axiom below for K. Basically, think of SF as doing for sensing actions what $Poss$ does for preconditions.

Figure 4.6 captures the idea that we would like s' to remain compatible with $s = S_0$, but s'' is to be discarded as a possibility. This leads to the following *domain-independent* successor state axiom for K to be included in the foundational axioms:

$$K(s', do(a, s)) \equiv \exists s''[K(s'', s) \wedge s' = do(a, s'') \wedge Poss(a, s'') \wedge (SF(a, s'') \equiv SF(a, s))].$$

Fig. 4.6 The compatibility of
situations after performing a
sensing action. The value of f
at the actual world $s = S_0$ is 1

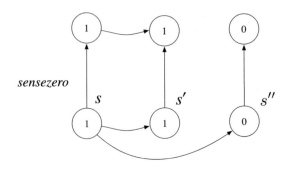

This says that if s'' is the predecessor of s', such that s'' was considered possible at s, then s' would be considered possible from $do(a, s)$ contingent on sensing outcomes.

We began with Fig. 4.4 where the initial situations disagree on the value of f, and consequently, the agent does not know f's value. After executing the sensing action, however, $do(sensezero, s'')$ is no longer accessible from $do(sensezero, s)$, because they disagree on the truth-status of SF. In contrast, $do(sensezero, s')$ continues to be accessible from $do(sensezero, s)$. The upshot is that the agent knows the value of f at $do(sensezero, s)$ and believes that this value is 1. In other words:

$$Knows(f = 1, do(sensezero, S_0)).$$

This is an instance of *knowledge expansion*. Likewise, suppose initially:

$$Knows(f = 0 \lor f = 1 \lor f = 2, S_0).$$

Then after sensing we have,

$$Knows(f = 1 \lor f = 2, do(sensezero, S_0)),$$

or equivalently, $Knows(f \neq 0, do(sensezero, S_0))$.

What about the sensing status of physical actions, and how do we ensure the accessibility relation is not affected? This can be achieved by simply letting SF to be vacuously true:

$$SF(incr(z), s) \equiv true.$$

As a result, if $K(s', s)$ then $K(do(a, s'), do(a, s))$ as well for physical actions by means of the same successor state axiom. In particular, as seen in Fig. 4.5, we have:

- $\neg Knows(f = 1, S_0)$; and
- $\neg Knows(f = 1, do(incr(1), S_0))$.

The agent remains ignorant about f's value after executions of $incr(z)$.

4.2.5   Axiomatization: Robot Sensing the Wall

Let us revisit the robot example from Fig. 3.2, and suppose we have a sonar sensor that tells the robot when its less than 4 units from the wall. In other words, we provide the following axiom:

$$SF(a, s) \equiv (a = sonar \wedge (h \leq 4)) \vee ((a = move \vee a = reverse) \wedge true).$$

So the physical actions provide a trivial sensing result, but for the sonar, SF is true only when $h \leq 4$.

Example 4.2   Let $\mathcal{D}$ be the union of the precondition and successor state axioms from Sect. 3.2.4, the above sensing axiom, and the following statement about the world and the agent's knowledge initially:

$$(h(S_0) = 5) \wedge \forall x. \neg Knows(h \neq x, S_0).$$

That is, the robot is 5 units away from the wall, but it does not have any idea how far it actually is. Let us also use the predicate $Close$ to abbreviate $h \leq 4$, to say that the robot is close to the wall.

We have the following entailments of $\mathcal{D}$:

1. $\neg Knows(Close, S_0)$.
   Clearly, the robot does not know it is close initially.
2. $\neg Knows(Close, do(move, S_0))$.
   Moving forward means the robot is within 4 units from the wall, but it does not know it.
3. $Knows(Close, do(sonar, do(move, S_0)))$.
   The sonar would inform the agent that $h \leq 4$, and so, it now knows that it is close.
4. $\neg Knows(Close, do(move, do(sonar, S_0)))$.
   Performing these actions in the reverse order does not mean that the robot will know its close. This is because performing the sensing action in S_0 will mean $\neg Close$ and so SF will be false. As far as the robot is concerned, h could be taking any value other than $\{0, \ldots, 4\}$. On moving forward, the robot only knows that its distance to the wall has decremented by one, whatever the initial value of h might have been.

4.2.6   Regression and Progression

Like with the classical variant, both regression and progression have been developed for the epistemic language, allowing one to eliminate actions for reasoning tasks.

The regression operator has an analogous property as seen in Sect. 8.1, reducing a formula of the form $Knows(\phi, do(\alpha, S_0))$ for some action sequence α to one of the form $Knows(\phi', S_0)$ which only mentions the situation term S_0.

Recall that previously regression worked by replacing a formula about what is true after an action by what must have been true before the action, by appealing to the RHS of successor state axioms. All of this is inherited for the epistemic variant. The key new feature is the use of the RHS of SF axioms when a sensing action is performed.

Example 4.3 Let us do an example, but gloss over the formal definition. Suppose we are attempting to check whether a slight variant of item 3 from above. Suppose we wish to check whether $\mathcal{D}$ entails

$$Knows(h \neq 6, do(sonar, do(move, S_0))).$$

On applying the regression operator, we have:

$$\mathcal{R}[Knows(h \neq 6, do(sonar, do(move, S_0)))] \tag{i}$$
$$= \mathcal{R}[Close(do(move, S_0)) \supset Knows(Close \supset (h \neq 6), (do(move, S_0)))] \wedge$$
$$\mathcal{R}[\neg Close(do(move, S_0)) \supset Knows(\neg Close \supset (h \neq 6), (do(move, S_0)))] \tag{ii}$$
$$= [(h \leq 5)[S_0] \supset Knows((h \leq 5) \supset (h \neq 6), S_0)] \wedge$$
$$[(h > 5)[S_0] \supset Knows((h > 5) \supset (h \neq 6), S_0)] \tag{iii}$$

Let us work through these steps. In step (i), we are regressing the agent knowing $h \neq 6$ after moving forwards and sensing. In step (ii), the sensing axioms come into play, and we are considering two possibilities. On the one hand, the RHS is true in the real world, in which case knowing the RHS implies the query. On the other, the RHS is false in the real world, in which case knowing that the RHS is false implies the query. The intuitive justification here is that on sensing, depending on how the world is, the agent knows of the outcome of the sensing which informs whatever it believes. Note that regression is purely syntactic, so it does not involve the initial theory but only the dynamic axioms. Regressing the formula in step (ii) w.r.t. the action of moving forward, we are left with the formula in step (iii). That is, for $h \leq 4$ to be true after moving forward, it must be that $h \leq 5$ is true initially. We have simplified $\neg(h \leq 5)$ as $h > 5$ for readability.

To now check whether the formula in step (iii) is entailed by the action theory, notice that since $(h = 5)$ is true initially, the antecedent in the first conjunct $(h \leq 5)$ is true. So for the formula in step (iii) to be true, all we then require is for $Knows((h \leq 5) \supset (h \neq 6), S_0)]$ to be entailed by the action theory. But the formula in the scope of $Knows$ is trivially true, and so the regressed formula is entailed by $\mathcal{D}$. It then follows that the query in step (i) is entailed by $\mathcal{D}$. Intuitively, because the agent has come to know the outcome of the sensing action, the query logically follows.

Example 4.4 To consider an example where the antecedent of the second conjunct in step (iii) would come out true, imagine the following query:

$$Knows(h \neq 6, do(sonar, S_0)).$$

This is like the formula in step (i), but the agent has not moved forward. So the regression of this leads to the formula in step (ii) but where the situation term everywhere is S_0 and not $do(move, S_0)$:

$$[Close(S_0) \supset Knows(Close \supset (h \neq 6), S_0)] \wedge [\neg Close(S_0)$$
$$\supset Knows(\neg Close \supset (h \neq 6), S_0)].$$

A further regression reduction does not happen, as there are no more actions to treat. Since the robot is not close initially, we get that $\mathcal{D}_0$ entails $\neg Close(S_0)$. But note also that the formula in the scope of $Knows$ is not trivially true, and neither is it entailed by $\mathcal{D}_0$. And thus, the query is not entailed by $\mathcal{D}$, and this is not surprising. The sensor only reveals information when the robot is close, and since the robot knew nothing about h initially, it will also not know that $h \neq 6$ after a sensing action.

An account of progression with knowledge can also be given. We omit the formal details of this account in its entirety here: it is fairly involved, and as with progression for the classical variant, uses second-order notions in general. To see an example:

Example 4.5 The progression of the initial theory from Sect. 4.2.5 w.r.t. the action sequence $[move \cdot sonar]$ would be:

$$(h = 4)[S_0] \wedge Knows(Close, S_0).$$

That is, the h value in the real world has reduced by a unit. If SF is true for the sensing action, the robot's knowledge includes the RHS of the sensing axiom. On the other hand, if SF is false, the robot's knowledge includes the negation of the RHS of the sensing axiom. Thus, there is clearly a parallel with step (ii) above.

Most of all, there is not much to be gained by looking at the formal definition as we will explore a first-order case when considering probabilities. (Moreover, our account is fairly different in form and style from the non-probabilistic versions.)

4.2.7 Knowledge-Based Programming

The high-level programming framework GOLOG, introduced in the previous chapter, allows one to construct routines using if-then-else and while constructs but where the conditions of these constructs are situation calculus formulas. This permits a novel way of designing agents, where we might interleave partial control specifications with planning. Note that the specifications could be further conditioned on what is true at runtime, by means of tests. Thus, the agent need not only rely on what the modeler might know at the time of building the agent.

With an explicit notion of knowledge, this idea is made richer. In place of a test program like "ϕ?", where it is implicit that ϕ refers to information available to the agent, we can explicitly check whether something is known, or not known to the agent. But perhaps the greatest benefit of an epistemic model is that we can now include sensing actions that complement what the agent initially knows with observations.

To appreciate this new feature, imagine writing a program to get the robot close to the wall with GOLOG as introduced in the previous chapter. Consider that the initial theory says nothing about h's initial value. Although we could execute a move if the robot is not close using:

$$\textbf{if } (h > 4) \textbf{ then } move \textbf{ endIf}$$

the program makes little sense. Since nothing is specified about h, the action is not performed in all worlds. That is, when looking at the entailments of such a program, there maybe some models of the action theory where the robot is close, and others where it is not. A forward action would only be performed in the latter. Moreover, how many move actions are sufficient to bring the robot close to the wall?

What is missing in the picture so as to specify the intended program with ease is the use of a sensor. Indeed, without the sensor, the robot would not know it is close to the wall regardless of how many moves it makes. This then leads *knowledge-based* program, where the test conditions can use *Knows*, and the sensor can be an atomic program. For example, our task of getting the robot close to the wall might be of the form:

$$\textbf{while } \neg Knows(Close) \textbf{ do } actsense \textbf{ endWhile}$$

where *actsense* might stand for the following sequential program:

$$move; sonar$$

This says that as long as the robot does not know it is close, it should execute a forward action followed by the action of sensing if its close. Recall that by executing the sensing action, the robot would know when it is close. If the robot is not yet close, the act-sense loop will repeat.

Notice that we have simply implemented a program based on the projection queries in the previous sections. If the robot is at the wall, performing the forward action might cause the robot to crash against the wall. The wiser strategy is perform a sensing acting first, and if it still the case that the robot is not at the wall, it should a move. For example, we could consider a program of the form:

$$\textbf{until } Knows(Close) \textbf{ do } senseact \textbf{ endUntil}$$

where *senseact* might stand for the following program:

$$sonar; safeact$$

and *safeact* might stand for acting only when the robot is not close:

if *Knows*(¬*Close*) **then** *move* **endIf**

Notice that by executing the sensing action, either *Knows*(*Close*) or *Knows*(¬*Close*); that is, the robot knows *whether* its close or not. This can be contrasted to complete ignorance, where ¬*Knows*(*Close*) ∧ ¬*Knows*(¬*Close*). The sensing action brings the robot to a different and more knowledgeable mental state than prior to that action, and of course, it is reasonable to only act after making sure the robot is not already close. All these subtleties are impossible to articulate in the non-epistemic variant where sensing actions cannot be utilized.

4.3 Notes

The formal semantics of knowledge owes its origin to Kripke [130, 131] and Hintikka [112], although epistemology is a much older field. Numerous communities have studied the use of epistemic logic for capturing systems, including game theory, distributed systems and, of course, artificial intelligence [73, 152]. For a study on the complexity of reasoning in epistemic logic, see the textbook by Fagin et al. [73]. That textbook also discusses common knowledge, the muddy children and the Byzantine general problem.

The use of situations as possible worlds leading to a language with reified worlds is due to Moore [171]. The further integration of Reiter's solution to the frame problem and regression is due to Scherl and Levesque [202]. Recently, [136] reconstruct the Scherl-Levesque scheme in a modal logic called $\mathcal{ES}$, which we will look to extend in a penultimate section with probabilities. For accounts of regression in the epistemic situation calculus (modal or otherwise), see [18, 136, 202]; for progression, see [19, 139, 162].

There are many other approaches to the integration of knowledge in a theory of action, including in dynamic logic [226]. See discussions on related efforts in [136, 190], for example.

Note that an explicit operator for knowledge makes reasoning about meta-knowledge possible [152]. Nested beliefs become even more prominent with the presence of multiple agents, where an agent might be reasoning about the knowledge and ignorance of another in service of collaborating or deceiving the other agent. Not surprisingly, this has use in areas such as game theory, distributed systems, among others [73]. Although we do not delve into multiple agents in this book, they are an exciting area of AI, and they bring new challenges to the framework we develop such as exchanging information (possibly over noisy channels) and jointly performing actions (such as lifting a heavy object with noisy actuators).

Reiter developed knowledge-based programming [190] in the situation calculus, although such programs were in consideration in other communities too [73]. For example, [104] look to characterize strategies in game theory using such programs. In [48], knowledge-based programs are also reconstructed in $\mathcal{ES}$.

In planning community, perhaps the best known equivalent of knowledge-based programming is the notion of *generalized planning* [149]. The key idea is to construct a plan structure that works for multiple initial situations, mainly differing in the value they ascribe to significant fluents. Analogous to the features of our robot example, a frequently studied problem is that of chopping down a tree of unknown thickness, equipped only with an actuator that cuts the tree and a sensor that only indicates when the tree is down [145]. Like in the robot example when an unknown number of forward actions may be necessary, the number of chops needed is not known. The sonar informing the robot when it is close to the wall is then the analogue to the sensor telling the agent when the tree is down. To capture the problem [199], we could entertain an explicit set of initial states (essentially an epistemic state)—although the use of the *Knows* modality and meta-knowledge is less common [208]—and use the sensor to discard possible states (much like SF axioms). Rather than attempting to synthesize a plan structure for a concrete set of initial states (say, where the tree thickness if 5 units but only that it is not known to the agent), we wish to have a plan structure that potentially works for arbitrary set of initial states parameterized over the tree thickness. A correct solution, then, is one that works for a tree of *any* thickness. The intent behind knowledge-based programs is similar. Since we do not know what the agents knows, the programs would need to be correct for all possible states of ignorance.

It is worth remarking that the preferred plan structure in generalized planning is a memoryless reactive automaton [21], which has a notable simplicity compared to the more complex theory needed for knowledge-based programming. This also likely makes them simpler for synthesis. See discussions in [14, 140] on the relative merits of the representations.

A very related area is that of *epistemic planning* [11, 111]. Here, we endeavor to plan in an epistemic language directly, in the sense of attempting to synthesize a sequence of actions that enable a goal mentioning epistemic modalities. This is arguably most interesting with multiple agents [172], where we are to find a plan that shares secrets among parties without being revealed to an adversary, for example.

Probabilistic Beliefs

5

> *On numerous occasions it has been suggested that the formalism*
> *[the situation calculus] take uncertainty into account by attaching*
> *probabilities to its sentences. We agree that the formalism will*
> *eventually have to allow statements about the probabilities of*
> *events, but attaching probabilities to all statements has the*
> *following objections:*
> *(i) It is not clear how to attach probabilities to statements*
> *containing quantifiers in a way that corresponds to the amount of*
> *conviction people have.*
> *(ii) The information necessary to assign numerical probabilities is*
> *not ordinarily available. Therefore, a formalism that required*
> *numerical probabilities would be epistemologically inadequate.*
> *—McCarthy and Hayes, Some philosophical problems from the*
> *standpoint of artificial intelligence*

Much of high-level AI research is concerned with the behavior of some putative agent, operating in a partially known environment. This agent is grappling with two fundamental sorts of reasoning problems. First, because the world is *dynamic*, it will need to reason about change: how its actions affect the state of the world. Pushing an object on a table may cause it to fall on the floor, where it will remain unless picked up. Second, because the world is *incompletely known*, the agent will need to make do with partial specifications about what is true. As a result, the agent will often need to augment what it believes about the world by performing perceptual actions, using sensors of one form or another. The previous two chapters accounted for these two problems in sequence.

For many AI applications, such as planning in the physical world and robotics, these reasoning problems are more involved. Here, it is not enough to deal with incomplete knowledge, where some formula ϕ might be unknown. One must also know which of ϕ or $\neg\phi$ is the more likely, and by how much. In addition, both the sensors and the effectors that the agent uses to modify its world are often subject to uncertainty in that they are noisy.

The current chapter explores the representation of this new type of complexity. Moreover, we address the matter in a way that pointedly tackles the concerns raised by McCarthy and Hayes.

© The Author(s), under exclusive license to Springer Nature Switzerland AG 2023   67
V. Belle, *Toward Robots That Reason: Logic, Probability & Causal Laws*,
Synthesis Lectures on Artificial Intelligence and Machine Learning,
https://doi.org/10.1007/978-3-031-21003-7_5

5.1   Beyond Knowledge and Deterministic Acting

Let us broadly consider the main question: how do we realize such a model of beliefs and actions? In the literature, there seem to be two disparate paradigms to capture the problem. At one extreme, there are logical formalisms, such as the situation calculus, which allows us to express strict uncertainty, and exploits regularities in the effects actions have on propositions to describe physical laws compactly. But all of this is largely limited to non-numerical uncertainty. At the other extreme, revising beliefs after noisy observations over error profiles is effortlessly addressed using probabilistic techniques such as Kalman filtering and dynamic Bayesian networks. However, in these frameworks, a complete specification of the dependencies between probabilistic variables is taken as given, in the form of the say, Bayesian network. This makes it difficult to deal with other forms of incomplete knowledge, such as disjunctive uncertainty, as well as complex actions that shift dependencies between variables in nontrivial ways.

As the quote above indicates, although rich first-order languages attempt to address important philosophical and representation problems raised by McCarthy and Hayes, and others, we will not want a pure probabilistic formalism. Such a formalism may force us to interpret things like disjunctions and existential quantifiers in a way that is notably different from how it understood in classical logic. We will also not want to be forced to provide numerical assessments for every statement, because this would make the formalism epistemologically inadequate in the sense suggested by McCarthy and Hayes.

What then is the emerging desiderata from these observations? Simply put, we want *all of first-order logic* and *probability theory*. In the desired formalism, it should be possible to specify a purely non-numerical account, recapturing the account of the previous chapter. Simultaneously, we will want to reason about probabilistic statements yielding the same conclusions afforded to us by probabilistic frameworks such as dynamic Bayesian networks. These two extremes will serve as a simple type of 'sanity check,' on the coherence of the formalism. Beyond that, of course, we will also want to mix and match aspects of the two paradigms: we might want to see the interplay between non-numerical initial knowledge and probabilistic error profiles in the actuators and sensors. Or perhaps we would want to see how probabilistic initial knowledge interacts with noisy sensors with a non-numerical error profile. The formalism should allow for such combinations.

Somewhat surprisingly, we show that a very natural extension to the epistemic model from the previous chapter, now with probabilities, leads to a formalism with exactly such properties. In classical possible-world semantics, a formula ϕ is believed to be true when ϕ comes out true in all possible worlds that are deemed accessible. In the extension, the degree of belief in ϕ is defined as a normalized sum over the possible worlds where ϕ is true of some nonnegative weights associated with those worlds. We could imagine inaccessible worlds are assigned a weight of zero. Like with the successor state axiom for K, the extension will include a successor state axiom regarding how the weight associated with a possible world changes as the result of acting and sensing. Previously where we had the specification of SF

axioms, we will now have the specification of axioms for the likelihood of action outcomes, including their failures. The properties of belief and belief change then emerge as a direct logical consequence of the initial constraints and these changes in weights.

The main advantage of this natural extension when compared to classical probabilistic frameworks is this: it allows a specification of belief that can be partial or incomplete, in keeping with whatever information is available about the application domain. It does not require specifying a prior distribution over some random variables from which posterior distributions are then calculated, as in Kalman filters. Nor does it require specifying the conditional independences among random variables and how these dependencies change as the result of actions, as in the dynamic and temporal extensions to Bayesian networks. In the situation calculus approach, some logical constraints are imposed on the initial state of belief. These constraints may be compatible with *one or very many initial distributions* and *sets of independence assumptions*. All the properties of belief will then follow at a corresponding level of specificity.

5.2   Weights and Likelihoods

To work through the representational subtleties, it will be helpful to motivate the subject matter by continuing to use the simple example of Fig. 3.2. The rest of the book will also use variations of this example.

For now, suppose the robot can move towards and away from the wall, and it is equipped with a distance sensor aimed at the wall. Assume the robot is a certain distance h from it. Here, the robot may not know the true value of h but may believe that it takes values from some set, say $\{1, 2, \ldots, 10\}$. Previously, the robot was only allowed to express its uncertainty about these values by means of a *disjunction:* $h = 1 \vee h = 2 \vee \ldots h = 10$. Now, the robot will be allowed to quantify this uncertainty.

Perhaps a very loose and unsophisticated comparison to draw to the disjunction is to say that all those values are *equally likely*. That is, as far as the robot is concerned, if we were to actually measure the distance between the robot and the wall, the numerical possibility of the distance being 1 unit is just as likely as it being 2 units, and just as likely as it being 3 units, and so on. This represents the robot's *degree of belief*, which consists of a *subjective* probabilistic assessment of the various values h can take. However, as we will discuss shortly, a disjunction at the logical level will also allow the robot to provide a different probabilistic assessment. For instance, the robot might believe, for whatever reason, values closer to zero are more likely than values further away from zero. So it might accord a subjective probability of 1/2 for $h = 1$, a probability of 1/4 for $h = 2$, and so on, such that the sum of the probabilities for the various values of h is always 1.

Beyond this initial situation, actions will also affect the degree of belief. If the sensor is noisy, a reading of, say, 5 units, does not guarantee that the agent is actually 5 units from the wall, although it should serve to increase the agent's degree of belief in that fact.

Analogously, if the robot intends to move by 2 units and the effector is noisy, it may end up moving by 2.12 units, which the agent does not get to observe. Be that as it may, the robot's degree of belief that it is closer to the wall should increase.

The idea here is that action outcomes are associated with probabilities. In the simplest instance, when an action is performed, there is a probability accorded to the action succeeding and enabling the change it was supposed to. This is complemented with the probability accorded to the same action failing. This is an *objective* probabilistic model of the possible outcomes, and with this knowledge, the robot then modifies its subjective view of the world appropriately after actions. Likewise, there is an *objective* probabilistic model of how the sensor fails, and with this knowledge, the robot modifies its subjective view based on the sensor reading. The overall picture is thus not dissimilar to the previous chapter, except in its numerical underpinning. From epistemic possibilities, we will now consider *weights* on worlds to numerically assess possibilities, together with *likelihood* functions to handle acting and sensing failures.

5.3   The Belief Macro

Like *Knows*, we will introduce a companion macro *Bel* for degrees of belief. Just as *Knows* expanded in terms of the epistemic fluent K, standing for the accessibility relation, we will now use a functional fluent p for the weights on that relation.

5.3.1   The Numeric Epistemic Fluent

The p fluent determines a probability distribution on situations, by associating situations with *weights*. More precisely, the term $p(s', s)$ denotes the relative *weight* accorded to situation s' when the agent happens to be in situation s. As one would for K, the properties of p in initial states, which vary from domain to domain, are specified with axioms as part of $\mathcal{D}_0$. For the robot example in Fig. 3.2, we might have:

$$\forall \iota, u.p(\iota, S_0)=u \equiv ((h(\iota)=2 \vee h(\iota) = 3) \wedge u = 0.5) \vee (h(\iota) \neq 2 \wedge h(\iota) \neq 3 \wedge u = 0).$$

This says that those initial situations where h has the integer values 2 or 3 obtain a weight of 0.5. All other situations, then, obtain 0 weight. Recall that in the non-numeric version, we would only be allowed to say:

$$\forall \iota. K(\iota, S_0) \equiv h(\iota) = 2 \vee h(\iota) = 3.$$

That is, only the initial situations where h takes the value 2 or takes the value 3 are epistemically possible from the actual world S_0, but it is not possible to say that they are equally likely, or perhaps that one value is more likely than the other.

We also might find it convenient to instead use the IF-THEN-ELSE notation and instead write:

$$p(\iota, S_0) = \text{IF} h(\iota) \in \{2, 3\} \text{THEN} 0.5 \text{ELSE} 0. \tag{5.1}$$

For very many standard distributions, we might further introduce a corresponding notation. Our initial beliefs could also be written as:

$$p(\iota, S_0) = \mathcal{U}(h(\iota); 2, 3).$$

In general, we use $p(\iota, S_0) = \mathcal{U}(h(\iota); a, b)$ to mean

$$p(\iota, S_0) = \text{IF} h(\iota) \in \{a, b\} \text{THEN} 1/(b - a + 1) \text{ELSE} 0.$$

That is, $\mathcal{U}$ is a discrete uniform distribution on the interval $[a, b]$. The same idea could be extended for a discrete approximation of a Gaussian, where, say, $\mathcal{N}(h(\iota); a, b)$ takes a to be the mean and b the standard deviation of the distribution.[1]

Note that because p is like any other fluent, the framework is more expressive than many probabilistic formalisms. For example, consider the following initial beliefs:

$$\forall \iota (p(\iota, S_0) = \mathcal{U}(h(\iota); 2, 3)) \vee \forall \iota (p(\iota, S_0) = \mathcal{N}(h[\iota]; 10, 20)).$$

This says that the agent believes h to be uniformly distributed on $[2, 3]$ or normally distributed with a mean of 10 and standard deviation of 20, without being able to say which.

The expressiveness is even more pronounced when we consider quantification. Just as writing such statements was easier to do with the *Knows* macro, than with the K fluent, we will now introduce the macro for degrees of belief. Note that this same macro applies to non-initial situations too, for which we will need to introduce additional concepts. But for now, let us interpret this only for the initial situation.

Definition 5.1 Suppose ϕ is any situation-suppressed formula. Then the degree of belief in ϕ is an abbreviation for:

$$Bel(\phi, s) \doteq \sum_{\{s':\phi[s']\}} p(s', s) \bigg/ \sum_{s'} p(s', s).$$

Here, the denominator is the *normalization factor* in the sense that it is the same expression as the numerator but with ϕ replaced by *true*. When $s = S_0$, the p-values of all situations s' relative to S_0 is taken into account for the sum term only if they satisfy ϕ. Then, *Bel* is the ratio of this total weight against the sum of p-values of all situations relative to S_0.

There are two constraints implicit in this definition. First, we would want p-values to be non-negative for the normalization factor to be well-defined, and more generally for *Bel* to

---

[1] Let us reiterate that we are considering discrete distributions only in the current chapter; the next chapter considers the capability to model continuous and mixed discrete-continuous distributions.

behave like a *probability distribution*. Second, just as it did not make sense to define non-initial situations accessible from S_0, it does not make sense for the p-values of non-initial situations to given non-zero values relative to S_0. These are enabled in the same style as one would expect in the situation calculus, via initial constraints and successor state axioms to be discussed below.

Using this macro, we might rephrase examples from above using formulas such as $Bel(h = 2, S_0) = 0.5$, for example.

Turning now to examples with quantifiers and quantifying-in, contrast the two sentences: $\exists x \, Bel(Alive(x), S_0) > 0$ versus $Bel(\exists x. \, Alive(x), S_0) > 0$. The former says there is some individual c, such that there is non-zero initial belief in c being alive. In other words, if we sum the p-values of initial situations where $Alive(c)$, it is non-zero. In contrast, the latter reflects the scenario where if we look at all the initial situations where somebody is alive, the sum of their p-values is non-zero. So, a different individual could be alive in each initial situation. Likewise, the formula $Bel(\forall x. \, Alive(x), S_0) > 0.9$ indicates that when we consider those worlds where every individual is alive, the sum of their p-values is more than 0.9.

Before turning to actions, it is worth noting that we could continue to use the relation K and *Knows* in terms of p:

$$K(s', s) \doteq p(s', s) > 0 \text{ and } Knows(\phi, s) \doteq Bel(\phi, s) = 1.$$

But to ensure that p is well-behaved for capturing distributions, we need to have p be non-negative. Moreover, just as K disallowed non-initial situations to be accessible from S_0, a similar constraint is needed for p. We lump these conditions below as an axiom that is assumed to be included in $\mathcal{D}_0$:

$$\forall \iota, s. \, p(s, \iota) \geq 0 \wedge (p(s, \iota) > 0 \supset Init(s)). \tag{P1}$$

Note that this is a stipulation about initial situations ι only. But we will also provide a *successor state axiom* for p, to be discussed shortly, that ensures that this constraint holds in all situations.

5.3.2 Likelihoods

Recall from our motivational example that with noisy sensing and acting, we have two types of complications:

- the intended action may not occur, and the value read on the sensor may be off from reality; and
- there is an associated likelihood with such failures.

In what follows, let us begin with the likelihood function and deal with the simpler case of noisy sensing. In the next chapter, when we further extend the account to deal with continuous probability distributions, we will discuss examples to sharpen our intuitions. This will then allow us to generalize the account for noisy acting. Incidentally, noisy acting will necessitate a slightly augmented language to allow for discrepancies between what was intended and what actually happened.

Let the term $l(a, s)$ be intended to denote the likelihood of action a occurring in situation s. This is perhaps best demonstrated using Fig. 3.2. Imagine a sonar aimed at the wall, which gives a reading for the true value of h. Supposing the sonar's readings are subject to additive Gaussian noise. That is, if now a reading of z were observed on the sonar, intuitively, those situations where $h = z$ should be considered more probable than those where $h \neq z$.[2] This occurrence is captured using likelihoods in the formalism. Basically, if $sonar(z)$ is the sonar sensing action with z being the value read, we specify a likelihood axiom describing its error profile as follows:

$$l(sonar(z), s) = u \equiv (z \geq 0 \wedge u = \mathcal{N}(z - h(s); \mu, \sigma^2)) \vee$$
$$(z < 0 \wedge u = 0). \tag{5.2}$$

This stipulates that the difference between a nonnegative reading of z and the true value h is normally distributed with a variance of σ^2 and mean of μ.

Clearly, the error profile of various hardware devices is application dependent, and it is this profile that is modeled as shown above using l. Notice, for example, when $\mu = 0$, which indicates that the sensor has no systematic bias, then $l(sonar(5), s)$ will be higher when $h(s) = 5$ than when $h(s) = 25$. Roughly, then, the idea is that after an observation, the weights on situations would get redistributed based on their compatibility with the observed value.

One may contrast such likelihood specifications to (trivial) ones for deterministic physical actions,[3] such as an action $move(z)$ of moving towards the wall by precisely z units. For such actions, we simply write

$$l(move(z), s) = u \equiv u = 1$$

in which case the p value of s is the same as that for $do(move(3), s)$. The idea that the weights of worlds are simply "transferred" to their successors is analogous to a philosophical notion called *imaging*.

Formally, we add *action likelihood* axioms to $\mathcal{D}$:

---

[2] As is standard in robotics, the reading observed is not in the control of the agent. Here, we assume that the value is given to us, and in that sense, the language is geared for projection queries. For example, we might be interested in the beliefs of the agent after obtaining a specific sequence of readings on the sonar. Integrating this language with an *online* framework that obtains such readings from an external source is addressed in a later chapter.

[3] Noisy actions will also involve non-trivial likelihood axioms.

Definition 5.2 Action likelihood axioms for each action type A are sentences of the form:

$$l(A(\vec{x}), s) = u \equiv \psi_A(\vec{x}, u, s).$$

Here $\psi_A(\vec{x}, u, s)$ is any formula characterizing the conditions under which action $A(\vec{x})$ has likelihood u in s.

In general, likelihood axioms can depend on any number of features of the world besides the fluent that the sensor is measuring. For example, imagine that the sonar's accuracy depends on the room temperature. We could then specify an error profile as follows:

$$
\begin{aligned}
l(sonar(z), s) = u \equiv \\
(z \geq 0 \wedge temp(s) \geq 0 \wedge u = \mathcal{N}(z - h(s); \mu, 1)) \vee \\
(z \geq 0 \wedge temp(s) < 0 \wedge u = \mathcal{N}(z - h(s); \mu, 16)) \vee \\
(z < 0 \wedge u = 0).
\end{aligned}
\tag{5.3}
$$

That is, the sonar's accuracy worsens severely when the temperature drops below 0, as seen by the larger variance.

We are now ready to provide the successor state axiom for p, which is analogous to the one for K:

$$
\begin{aligned}
p(s', do(a, s)) = u \equiv \\
\exists s'' \, [s' = do(a, s'') \wedge Poss(a, s'') \wedge u = p(s'', s) \times l(a, s'')] \\
\vee \neg \exists s'' \, [s' = do(a, s'') \wedge Poss(a, s'') \wedge u = 0].
\end{aligned}
\tag{P2}
$$

This says that the weight of situations s' relative to $do(a, s)$ is the weight of their predecessors s'' times the likelihood of a contingent on the successful execution of a at s''.

5.3.3 Axiomatization: The One-Dimensional Robot

To see an application of this axiom using the specifications (5.1) and (5.2), consider two situations s and s' associated with the same weight initially, as shown in Fig. 5.1. These situations have h values 2 and 3 respectively. Suppose the robot obtains a reading of 2 on the sensor. Given a sensor with mean $\mu = 0$ and variance $\sigma^2 = 1$, the likelihood axiom is such that the weight of the successor of s is higher than that of s' because the h value at s coincides with the sensor reading. More precisely, the weight for the successor of s is given by the prior weight 0.5 multiplied by the likelihood factor $\mathcal{N}(z - h(s); \mu, \sigma^2) = \mathcal{N}(2 - 2; 0, 1) = \mathcal{N}(0, 0, 1)$. The weight for the successor of s' is obtained analogously.

Interestingly, by means of the above axiom, if predecessors are not *epistemically* related, which is another way of saying that $p(s'', s)$ is 0, then their successors will also not be. Similarly, when a is not executable at s'', its successor is no longer accessible from $do(a, s)$. One other consequence of (P1) and (P2) is that $(p(s', s) > 0)$ will be true only when s' and

Fig. 5.1 Situations with accessibility relations after noisy sensing. The numbers inside the circles denote the h values at these situations. Dotted circles denote a lower weight in relation to their epistemic alternatives

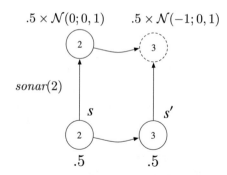

s share the same history of actions. This is because (P1) insists that initial situations are only epistemically related to other initial ones, and (P2) respects this relation over actions.

Looking back at our definition for Bel, we see that it is perfectly reasonable to obtain the degree of belief in ϕ at s by looking at $p(s', s)$ over all situations s': clearly, if s' and s do not share histories, the p value will be zero anyway.

Example 5.3 Given a basic action theory consisting of (P1), (P2), (5.1) and (5.2), it is not hard to show that would entail the following:

1. $Bel(h = 4, S_0) = 0$.
   Since the agent believes h is uniformly distributed on $\{2, 3\}$, it accords a degree of belief of 0 to any other value for h.
2. $Bel(h = 3 \lor h = 4, S_0) = 0.5$.
   The degree of belief in the disjunction ($h = 3 \lor h = 4$) is non-zero because the degree of belief in one of the disjuncts in non-zero.
3. $Bel(h = 2, do(sonar(2), S_0)) > Bel(h = 3, do(sonar(2), S_0)) > 0$.
   As explained in Fig. 5.1, despite the sensor being noisy, obtaining a reading of 2 on the sonar would mean h taking a value of 2 will now be considered more likely than it having the value of 3. Unlike the categorical version, worlds are not discarded and so the belief in $h = 3$ after performing a sensing action is not zero.[4]

The next chapters will develop more elaborate examples using an extended account with continuous distributions and noisy acting.

---

[4] This is mainly owing to the sensor's error profile being (5.2). It is possible to imagine sensors that discard all values other than the true value (i.e., an exact sensor), or one that discards, say, all values higher than the read value. In the latter case, on reading a value of 2 on the sensor, $h = x$ for all $x > 2$ would be accorded a degree of belief of 0. We will discuss such error profiles in the penultimate chapter.

5.4   Notes

This chapter focused on degrees of belief in a first-order dynamic setting, drawn from the work of BHL [8],[5] and its refinement by the author and his collaborators [22, 28]. Related efforts on belief update via sensor information can be broadly classified into two camps: the literature on probabilistic formalisms, and those that extend logical languages. We discuss them in turn. At the outset, we remark that the focus of our work is on developing a general framework, and not on computational considerations, efficiency or otherwise. As mentioned, the next chapter will discuss the continuous extension; in particular, the existing scheme of BHL is generalized to handle both discrete and continuous probability distributions while retaining all the advantages. In anticipation of that, we take the liberty of also mentioning related advances, where applicable, of modelling continuous distributions. The discussion below recaps threads raised in Chap. 2 and others, but it is perhaps useful to revisit the differences to related efforts now that the formal machinery of our proposal has been studied in more detail.

From the perspective of probabilistic modeling, graphical models [124], such as Bayesian networks [178], are important formalisms for dealing with probabilistic uncertainty in general, and the uncertainty that would arise from noisy sensors in particular [219]. Mainly, when random variables are defined by a set of dependencies, the density function can be compactly factorized using these formalisms. The significance of such formalisms is computational, with reasoning methods, such as filtering, being a fundamental component of contemporary robotics and machine learning technologies [60, 78, 219]. On the representation side, however, these formalisms have difficulties handling strict uncertainty, as would arise from connectives such as disjunctions. (Proposals such as credal networks [52] allow for certain types of partial specifications, but still do not offer the generality of arbitrary logical constraints.) The last decade has also seen the use of logic-based solvers for reasoning in such networks [43], primarily based on propositional satisfiability technology. While these solvers do allow the inclusion of logical constraints in addition to network dependencies, they should be thought of imposing hard constraints on the distribution, as opposed to entertaining, say, multiple distributions that may be compatible with a disjunction.

Moreover, since rich models of actions are rarely incorporated, shifting conditional dependencies and distributions are hard to address in a general way. While there are graphical model frameworks with an account of actions, such as [54, 98], they too have difficulties handling strict uncertainty and quantification. To the best of our knowledge, no existing probabilistic formalism handles changes in state variables like those considered here.

This inherent limitation of probabilistic formalisms led to a number of influential proposals on combining logical and probabilistic specifications [174]. (The synthesis of deductive reasoning and the probability calculus has a long and distinguished history [39, 82] that we do not review here; see [101] and references therein.) The works of Bacchus and Halpern

---

[5] See [154] for a discussion on imaging.

[5, 72], for example, provide the means to specify probabilities about the domain together with probabilities about propositions; the former is appropriate for statistical information such as "the probability that a typical male from this population suffers from obesity is 0.7", whereas the latter, as in this book, is appropriate for degrees of belief. Many such (early) logics did not explicitly address reasoning about actions. As we noted, treating actions in a general way requires, among other things, addressing the frame problem, reasoning about what happened in the past and projecting the future, handling contextual effects, as well as appropriate semantical machinery. We piggybacked on the powerful situation calculus framework, and extended that theory for reasoning about uncertainty.

Recently in AI, limited versions of probabilistic logics have been discussed, in the form of relational graphical models, including Markov logic networks, probabilistic databases and probabilistic programming [68, 89, 125, 169, 184, 188, 193, 205, 210]. Some have been further extended for continuous probability distributions and temporal reasoning [2, 29, 45, 177]. Overall, these limit the first-order expressiveness of the language, do not treat actions in a general way, and do not handle strict uncertainty. Admittedly, syntactical restrictions in these frameworks are by design, in the interest of tractability (or at least decidability) w.r.t. inference, as they have origins in the richer probabilistic logical languages mentioned above [5, 72]. From the point of view of a general-purpose representation language, however, they are lacking in the kinds of features that we emphasize here.

From a modal logic viewpoint, the interaction between categorical knowledge, on the one hand, and degrees of belief, on the other, is further discussed in [72, 109, 170]. While these are essentially propositional, there are first-order variants too [20]. But to be closer in spirit to our work here, we need to look at knowledge representation languages for reasoning about action and knowledge, which we refer to as action logics.

The situation calculus [167, 190], which has been the sole focus of this book, is one such language. There are others, of course, such as the event calculus [127], dynamic logic [227, 228], the fluent calculus [218], and formalisms based on the stable model semantics [86].

As discussed previously, in the situation calculus, a monotonic solution to the frame problem was provided by Reiter [189]. The situation calculus was extended to reason about knowledge whilst incorporating this solution in [202], and to reason about noisy effectors and sensors by BHL.[6] Other action logics have enjoyed similar extensions. For example, [126] proposes an extension to dynamic logic for reasoning about degrees of belief and noisy actions, and [12] provides a computational framework for probabilistic reasoning using the stable model semantics, but they are propositional. In [164, 215], the fluent calculus was extended for probabilistic beliefs and noisy actions. None of these admit continuous probability distributions.

---

[6] Throughout, we operated under the setting of knowledge expansion, that is, observations are assumed to resolve the agent's uncertainty and never contradict what is believed. The topic of *belief revision* lifts this assumption [1], but it is not considered here. See [203] for an account of belief revision in the situation calculus.

The situation calculus has also been extended for uncertainty modeling in other directions. For example, [165] consider discrete noisy actions over complete knowledge, that is, no degrees of belief. In later work, [80] treat continuous random variables as meta-linguistic functions, and so their semantics is not provided in the language of the situation calculus. This seems sufficient for representing things like products of probabilistic densities, but it is not an epistemic account in a logical or probabilistic sense. A final prominent extension to the situation calculus for uncertainty is the embedding of decision-theoretic planning in GOLOG [33, 34]. Here, actions are allowed to be nondeterministic, but the assumption is that the actual state of the world is fully observable. (It essentially corresponds to a fully observable Markov decision process [32].) In this sense, the picture is a special case of the BHL framework. It is also not developed as a model of belief. While this line of work has been extended to a partially observable setting [196], the latter extension is also not developed as a model of belief. Perhaps most significantly, neither of these support continuous probability distributions, nor strict uncertainty at the level of probabilistic beliefs.

It is worth noting that real-valued fluents in action logics turn out to be useful for modeling resources and time. See, for example, [79, 96, 110]. These are, in a sense, complementary to an account of belief.

Outside the logical literature, there are a variety of formalisms for modeling noisy dynamical systems. Of these, partially observable Markov decision processes (over discrete and continuous random variables) are perhaps the most dominant [219]. They can be seen as belonging to the literature on probabilistic planning languages [133, 232]. Recent probabilistic planning languages [195], moreover, combine continuous Bayesian networks and classical planning languages. Planning languages, generally speaking, only admit a limited set of logical connectives, constrain the language for specifying dynamic laws (that is, they limit the syntax of the successor state axioms), and do not handle strict uncertainty.

In sum, the account in this and the next chapter chapter is a simple yet general approach to handle degrees of belief, noisy sensors and effectors over discrete and continuous probability distributions in a general way. The proposal allows for partial and incomplete specifications, and the properties of belief will then follow at a corresponding level of specificity.

Continuous Distributions

6

> *We have to continually be jumping off cliffs and developing our*
> *wings on the way down.*
> —***Kurt Vonnegut**, If This Isn't Nice, What Is?*

The model of belief from the previous chapter is very general in extending the expressive-ness of the situation calculus to allow for probabilistic specifications, however partial or incomplete they may be. However, it does have one serious drawback: it is ultimately based on the addition of weights and is therefore restricted to fluents having discrete finite values. In fact, this is true for most logic-based formalisms. This is in stark contrast to robotics and machine learning applications, where event and observation variables are characterized by continuous distributions, or perhaps combinations of discrete and continuous ones. There is no way to in the language from the previous chapter that the initial value of h is any real number drawn from a uniform distribution on the interval [2, 12]. One would again expect the belief in $(h < 9)$ to be 0.7, but instead of being the result of *summing* weights, it must now be the result of *integrating* densities over a suitable space of values.

So, on the one hand, our account and others like it can be seen as general formal theories that attempt to address important philosophical problems such as those raised by McCarthy and Hayes above. But on the other, a serious criticism levelled at this line of work, and indeed at much of the work in reasoning about action, is that the theory is far removed from the kind of probabilistic uncertainty and noise seen in typical robotic applications.

The goal of this chapter is to show how with minimal additional assumptions this serious limitation can be lifted. By lifting this limitation, one obtains, for the first time, a logical language for representing real-world robotic specifications without any modifications, but also extend beyond it by means of the logical features of the underlying framework. In particular, we present a *formal specification* of the degrees of belief in formulas with real-valued fluents (and other fluents too), and how belief changes as the result of acting and sensing. Our account will retain the advantages of the model from the previous chapter but

© The Author(s), under exclusive license to Springer Nature Switzerland AG 2023
V. Belle, *Toward Robots That Reason: Logic, Probability & Causal Laws*,
Synthesis Lectures on Artificial Intelligence and Machine Learning,
https://doi.org/10.1007/978-3-031-21003-7_6

work seamlessly with discrete probability distributions, probability densities, and perhaps most significantly, with difficult combinations of the two. More broadly, we believe the model of belief proposed in this work provides the necessary bridge between logic-based reasoning modules, on the one hand, and probabilistic specifications as seen in real-world data-intensive applications, on the other.

6.1   Belief Reformulated

The definition for degrees of belief from the previous chapter is intuitive and simple. It is closely fashioned after the semantics for belief in modal probability logics, where the probabilities of formulas is calculated from the weights of possible worlds satisfying the formula. Unfortunately, this definition is not easily amenable to generalizations. Notice, for example, that *Bel* is well-defined only when the sum over all those situations s' such that $\phi[s']$ holds is finite. This immediately precludes domains that involve an infinite set of situations agreeing on a formula. Moreover, the definition does not have an obvious analogue for continuous probability distributions. Observe that such an analogue would involve *integrating* over the space of situations, which makes little sense. Indeed, it is not certain what the space of situations would look like in general, but even if this was fixed, how such a thing can be tinkered with so as to obtain an appropriate notion of *integration* is far from obvious.

Therefore, what we propose is to shift the calculating of probabilities from situations to *fluent values*, that is, to the well-understood domain of numbers. The current section is an exploration of this idea. What we will show in this section is that our definition of *Bel* can be reformulated as a summation over numeric indices. That will allow, among other things, a seamless generalization from summation to integration, which is to be the topic of the next section.

To prepare for that, in addition to IF-THEN-ELSE and case statements, we will introduce another kind of conditional term for convenience. This involves a quantifier and a default value of 0, like in formula (P2). If z is a variable, ψ is a formula and t is a term, we use $\langle z.\psi \rightarrow t \rangle$ as a logical term characterized as follows:

$$\langle z.\psi \rightarrow t \rangle = u \doteq [(\exists z\psi) \supset \forall z(\psi \supset u = t)] \wedge [(\neg\exists z\psi) \supset u = 0)].$$

The notation says that when $\exists z\psi$ is true, the value of the term is t; otherwise, the value is 0. When t uses z (the usual case), this will be most useful if there is a *unique z* that satisfies ψ.

Returning to the task at hand, we will now need a way to enumerate the primitive fluent terms of the language. Intuitively, these correspond to the probabilistic variables in the language. Perhaps the simplest way is to assume there are n fluents $f_1, f_2, \ldots, f_n$ in the language which take no arguments other than the situation argument, and that they take their values from some finite sets. We can then rephrase our prior abbreviation for *Bel* as follows:

Definition 6.1 Suppose ϕ is a situation-suppressed formula. Let $Bel(\phi, s)$ be an abbreviation for[1]:

$$\frac{1}{\gamma} \sum_{\vec{x}} \sum_{s'} \begin{cases} p(s', s) & \text{if } \bigwedge f_i(s') = x_i \wedge \phi[s'] \\ 0 & \text{otherwise} \end{cases}$$

Here and elsewhere, γ is the numerator but with ϕ replaced by *true*. This can be seen to also apply to Definition 5.1.

Definition 6.1 suggests that for each possible value of the fluents, we are to sum over all possible situations and for each one, if the fluents have those values and ϕ holds, then the p value is to be used, and 0 otherwise.[2] Roughly speaking, if one were to group situations satisfying $\bigwedge f_i(s) = x_i$ into sets for every possible vector $\vec{x}$, the union of these sets would give the space of situations. Our claim about the relationship between the two abbreviations can be made precise as follows:

Theorem 6.2 *Let $\mathcal{D}$ be any basic action theory and ϕ any formula. Then the abbreviations for $Bel(\phi, s)$ from Definitions 5.1 and 6.1 define the same number.*

Be that as it may, Definition 6.1 still involves summations over situations. To arrive at a definition that eschews the summing of situations, we start with the case of initial situations. In this matter, we will be insisting on a precise space of initial situations. For this, let us consider an axiomatization of the situation calculus for multiple initial situations from Levesque, Pirri and Reiter. This includes a sentence saying there is precisely one initial situation for *any* possible vector of fluent values. This can be written as follows:

$$[\forall \vec{x} \, \exists \iota \bigwedge f_i(\iota) = x_i] \wedge [\forall \iota, \iota'. \bigwedge f_i(\iota) = f_i(\iota') \supset \iota = \iota'] \tag{P3}$$

Recall that i ranges over the indices of the fluents, that is, $\{1, \ldots, n\}$. Under the assumption (P3), we can rewrite Definition 5.1 for $s = S_0$ as

$$Bel(\phi, S_0) \doteq \frac{1}{\gamma} \sum_{\vec{x}} \langle \iota. \bigwedge f_i(\iota) = x_i \wedge \phi[\iota] \rightarrow p(\iota, S_0) \rangle \tag{B0}$$

---

[1] For readability, we often drop the index variables in sums and connectives when the context makes it clear: in this case, i ranges over the set $\{1, \ldots, n\}$, that is, the indices of the fluents the language.

[2] Essentially, functional fluents are assumed to not take any object arguments. More generally, if we assume that the arguments of k-ary fluents are drawn from finite sets, an analogous enumeration of ground functional fluent terms is possible. Understandably, from the point of view of situation calculus basic action theories, where fluents are also usually allowed to take arguments from *any* set, including infinite ones, this is a limitation. But in probabilistic terms, this would correspond to having a joint probability distribution over infinitely many, perhaps uncountably many, random variables. We know of no general logical account of this sort, so we leave the matter as is.

The two abbreviations, in fact, are equivalent:

Theorem 6.3 *Let $\mathcal{D}$ be any basic action theory, ϕ as before, and suppose $\mathcal{D}_0$ includes (P3). Then the abbreviations for $Bel(\phi, S_0)$ in Definition 5.1 and (B0) define the same number.*

This shows that for S_0, summing over possible worlds can be replaced by summing over fluent values.

Unfortunately, (B0) is only geared for initial situations. For non-initial situations, the assumption that no two agree on all fluent values is untenable. To see why, imagine an action $move(z)$ that moves the robot z units to the left (towards the wall) but that the motion stops if the robot hits the wall. The successor state axiom for fluent h, then, might be like this:

$$h(do(a, s)) = u \equiv$$
$$\neg \exists z (a = move(z)) \wedge u = h(s) \ \vee \tag{6.1}$$
$$\exists z (a = move(z) \wedge u = \max(0, h(s) - z)).$$

In this case, if we have two initial situations that are identical except that $h = 3$ in one and $h = 4$ in the other, then the two distinct successor situations that result from doing $move(4)$ would agree on all fluents (since both would have $h = 0$). Ergo, we cannot sum over fluent values for non-initial situations unless we are prepared to count some fluent values more than once.

It turns out there is a simple way to circumvent this issue by appealing to Reiter's solution to the frame problem. Indeed, Reiter's solution gives us a way of computing what holds in non-initial situations in terms of what holds in initial ones, which can be used for computing belief at arbitrary successors of S_0. More precisely,

Definition 6.4 *(Degrees of belief (reformulated).)* Let ϕ be any formula. Given any sequence of ground action terms $\alpha = [a_1, \ldots, a_k]$, let

$$Bel(\phi, s) \doteq \frac{1}{\gamma} \sum_{\vec{x}} P(\vec{x}, \phi, s)$$

where if $s = do(\alpha, S_0)$ then

$$P(\vec{x}, \phi, do(\alpha, S_0)) \doteq \langle \iota. \bigwedge f_i(\iota) = x_i \ \wedge \ \phi[do(\alpha, \iota)] \ \rightarrow \ p(do(\alpha, \iota), do(\alpha, S_0)) \rangle.$$

To say more about how (and why) this definition works, we first note that by (P1) and (P2), p will be 0 unless its two arguments share the same history. So the s' argument of p in Definition 5.1 is *expanded* and written as $do(\alpha, \iota)$ in Definition 6.4. By ranging over all fluent values, we range over all initial ι as before, but without ever having to deal with fluent values in non-initial situations. Of course, we test that the ϕ holds and use the p weight in the appropriate non-initial situation. In particular, owing to p's successor state axiom (P2), the weight for non-initial situations accounts for the likelihood of actions executed in the history. We now establish the following result:

Theorem 6.5 *Let* $\mathcal{D}$ *be any basic action theory with (P3) initially,* ϕ *any formula, and* α *any sequence of ground actions terms. Then the abbreviations for* $Bel(\phi, do(\alpha, S_0))$ *in Definitions 5.1 and 6.4 define the same number.*

Thus, by incorporating a simple constraint on initial situations, we now have a notion of belief that does not require summing over situations.

Readers may notice that our reformulation only applies when we are given an explicit sequence α of actions, including the sensing ones. But this is just what we would expect to be given for the projection problem, where we are interested in inferring whether a formula holds after an action sequence. In fact, we can use regression on the ϕ and the p to reduce the belief formula from Definition 6.4 to a formula involving initial situations only. A subsequent chapter discusses this.

6.2 From Weights to Densities

The framework presented so far is *fully discrete*, which is to say that fluents, sensors and effectors are characterized by finite values and finite outcomes. Belief in ϕ, in particular, is the summing over a finite set of situations where ϕ holds. We now generalize this framework. We structure our work by first focusing on *fully continuous domains*, which is to say that fluents, sensors and effectors are characterized by values and outcomes ranging over $\mathbb{R}$. This section, in particular, explores the very first installment: effectors are assumed to be deterministic, but *sensors* have continuous noisy error profiles. A subsequent section, then, allows both effectors and sensors to have continuous noisy profiles.

Let us begin by observing that the uncountable nature of continuous domains precludes summing over possible situations. In this section, we present a new formalization of belief in terms of *integrating* over fluent values. This, in particular, is made possible by the developments in the preceding section.

Allowing real-valued fluents implies that there will be uncountably many initial situations. Imagine, for example, the scenario from Fig. 3.2, and that the fluent h can now be any nonnegative real number. Then for any nonnegative real x there will be an initial situation where $(h = x)$ is true. Suppose further that $\mathcal{D}_0$ includes:

$$p(\iota, S_0) = \begin{cases} 0.1 & \text{if } 2 \leq h(\iota) \leq 12 \\ 0 & \text{otherwise} \end{cases} \tag{6.2}$$

which says that the true value of h initially is drawn from a uniform distribution on the interval [2,12]. Then there are uncountably many situations where p is non-zero initially.

So the p fluent now needs to be understood as a *density*, not as a weight. (That is, we now interpret $p(s', s)$ as the *density* of s' when the agent is in s.) In particular, for any x, we would expect the initial degree of belief in the formula $(h = x)$ to be 0, but in $(h \le 12)$ to be 1.

When actions enter the picture, even if deterministic, there is more to be said. Numerous subtleties arise with p in non-initial situations. For example, if the robot were to do a *move*(4) there would be an uncountable number of situations agreeing on $h = 0$: namely, those where $2 \le h \le 4$ was true initially. In a sense, the point $h = 0$ now has *weight*, and the degree of belief in $h = 0$ should be 0.2. On the other hand, the other points $h \in (0, 8]$ should retain their densities. That is, belief in $h \le 2$ should be 0.4 but belief in $h = 2$ should still be 0. In effect, we have moved from a continuous to a *mixed* distribution on h. Of course, a subsequent rightward motion will retain this mixed density. For example, if the robot were to now move away by 4 units, the belief in $h = 4$ would then be 0.2.

To address the concern of belief change in continuous domains, we now present a generalization. One of the advantages of our approach is that we will not need to specify how to handle changing densities and distributions like the ones above. These will emerge as side-effects, that is, shifting density changes will be *entailed* by the action theory.

For our formulation of belief, we first observe that we have fluents $f_1, \dots, f_n$ as before, that take no argument other than the situation term but which now take their values from $\mathbb{R}$. Then:

Definition 6.6 (*Degrees of belief (continuous noisy sensors).*) Let ϕ be any situation-suppressed formula, and $\alpha = [a_1, \dots, a_k]$ any ground sequence of action terms. The *degree of belief* in ϕ at s is an abbreviation:

$$Bel(\phi, s) \doteq \frac{1}{\gamma} \int_{\vec{x}} P(\vec{x}, \phi, s)$$

where, as in Definition 6.4, if $s = do(\alpha, S_0)$ then

$$P(\vec{x}, \phi, do(\alpha, S_0)) \doteq \langle \iota . \bigwedge f_i(\iota) = x_i \wedge \phi[do(\alpha, \iota)] \rightarrow p(do(\alpha, \iota), do(\alpha, S_0)) \rangle.$$

That is, the belief in ϕ is obtained by ranging over all possible fluent values, and integrating the densities of situations where ϕ holds. If we were to compare the above definition to Definition 6.4, we see that we have simply shifted from summing over finite domains to integrating over reals. In fact, we could read P as the *(unnormalized)* density associated with a situation satisfying ϕ. As discussed, by insisting on an explicit world history, the ι need only range over initial situations, giving us the exact correspondence with fluent values.

This completes our new definition of belief. To summarize, our extension to the previous scheme is defined using a few convenient abbreviations, such as for *Bel* and mathematical integration, and where an action theory consists of:

1. $\mathcal{D}_0$ (with (P1)) as usual, but now also including (P3);
2. precondition axioms as usual;
3. successor state axioms, including one for p, namely (P2), as usual;
4. foundational domain-independent axioms as usual; and
5. action likelihood axioms, one for each action type.

Note that, apart from (P3) and *Bel*'s new abbreviation, we carry over precisely the same components as before. By and large, the extension, thus, retains the simplicity of their proposal, and comes with minor additions. We will show that it has reasonable properties using an example and its connection to Bayesian conditioning below.

In the sequel, we assume, without explicitly mentioning so, that basic action theories include the sentences (P1), (P2) and (P3).

6.3 Bayesian Conditioning

We now explicate the relationship between our definition for *Bel* and *Bayesian conditioning*. Bayesian conditioning is a standard model for belief change w.r.t. noisy sensing and it rests on two significant assumptions. First, sensors do not physically change the world, and second, conditioning on a random variable f is the same as conditioning on the event of observing f.

In general, in the language of the situation calculus, there need not be a distinction between sensing actions and physical actions. In that case, the agent's beliefs are affected by the sensed value as well as any other physical changes that the action might enable to adequately capture the "total evidence" requirement of Bayesian conditioning.

The second assumption expects that sensors only depend on the true value for the fluent. For example, in the formulation of (5.2) the sonar's error profile is determined solely by h. But to suggest that the error profile might depend on other factors about the environment, as formulated by (5.3) for example, goes beyond this simplified view. In fact, here, the agent also learns about the room temperature, apart from sensing the value of h.

Thus, our theory of action admits a view of dynamical systems far richer than the standard setting where Bayesian conditioning is applied. Be that as it may, when a similar set of assumptions are imposed as axioms in an action theory, we obtain a sensor fusion model identical to Bayesian conditioning. We state the property formally for continuous variables below.

We begin by stipulating that actions are either of the *physical* type or of the *sensing* type, the latter being the kind that do not change the value of any fluent, that is, such actions do not appear in the successor state axioms for any fluent. Now, if $obs(z)$ senses the true value of fluent f, then assume the sensor error model to be:

$$l(obs(z), s) = u \equiv u = Err(z, f(s))$$

where $Err(u_1, u_2)$ is some expression with only two free variables, both numeric. This captures the notion established above: the error model of a sensor measuring f depends only on the true value of f, and is independent of other factors. Finally, for simplicity, assume *obs(z)* is always executable:

$$Poss(obs(z), s) \equiv true.$$

Then we obtain:

Theorem 6.7 *Suppose $\mathcal{D}$ is any basic action theory with likelihood and precondition axioms for obs(z) as above, ϕ is any formula mentioning only f, and $u \in \{x_1, \ldots, x_n\}$ that f takes a value from. Then we obtain*:

$$\mathcal{D} \models Bel(\phi, do(obs(z), S_0)) = \frac{\int_{\vec{x}}[P(\vec{x}, \phi \wedge f = u, S_0) \times Err(z, u)]}{\int_{\vec{x}}[P(\vec{x}, f = u, S_0) \times Err(z, u)]}$$

That is, the *posterior belief* in ϕ is obtained from the *prior density* and the error likelihoods for all points where ϕ holds given that z is observed, normalized over all points.

The usual case for posteriors are formulas such as $b \leq f \leq c$, which is estimated from the prior and error likelihoods for all points in the range $[b, c]$, as demonstrated by the following consequence:

Corollary 6.8 *Suppose D is any basic action theory with likelihood and precondition axioms for $obs(z)$ as above, f is any fluent, and u is a variable from $\vec{x} = \langle x_1, \ldots, x_m \rangle$ that f takes a value from. Then we obtain*:

$$\mathcal{D} \models Bel(a \leq f \leq b, do(obs(z), S_0)) = \frac{\int_{\vec{x}}[P(\vec{x}, f = u \wedge a \leq u \leq b, S_0) \times Err(z, u)]}{\int_{\vec{x}}[P(\vec{x}, f = u, S_0) \times Err(z, u)]}.$$

More generally however, and unlike many probabilistic formalisms, we are able to reason about any logical property ϕ of the random variable f being measured.

6.4   Axiomatization: A Two-Dimensional Robot

Using an example, we demonstrate the formalism and Theorem 6.7 in particular. To reason about the beliefs of our robot, let us build a simple basic action theory $\mathcal{D}$. We extend the setting from Fig. 3.2 to a 2-dimensional grid, as shown in Fig. 6.1. As before, let h be the fluent denoting its horizontal position (that is, its distance to the wall), and let the robot's vertical position be given by a fluent v. The components of $\mathcal{D}$ are as below.

Fig. 6.1 A robot in a
2-dimensional grid

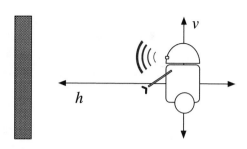

- Imagine a *p*-axiom of the form:

$$p(\iota, S_0) = \begin{cases} 0.1 \times \mathcal{N}(v(\iota); 0, 16) & \text{if } 2 \leq h(\iota) \leq 12 \\ 0 & \text{otherwise} \end{cases} \tag{6.3}$$

This says that the value of *v* is normally distributed about the horizontal axis with variance 16, and independently, that the value of *h* is uniformly distributed between 2 and 12. Note also that initial beliefs can be specified for $\mathcal{D}_0$ using *Bel* directly. For example, to express that the true value of *h* is believed to be uniformly distributed on the interval [2, 12] we might equivalently include the following theory in $\mathcal{D}_0$:

$$\{Bel(2 \leq h \leq 12, S_0) = 0.1, Bel(h \leq 2 \vee h \geq 12, S_0) = 0\},$$

and analogously for the fluent *v*.
For this example, a simple distribution has been chosen for illustrative purposes. In general, recall from the previous chapter that the *p*-specification does not require the variables to be independent, nor does it have to mention all variables.
- For simplicity, let us assume that actions are always executable, i.e., that $\mathcal{D}$ includes

$$Poss(a, s) \equiv true \tag{6.4}$$

for all actions *a*. For this example, we assume three action types: action *move(z)* that moves the robot *z* units towards the wall, action *up(z)* that moves the robot *z* units away from the horizontal axis, and action *sonar(z)* that gives a reading of *z* for the distance between the robot and the wall.
- The successor state axiom for *h* is as in (6.1), and the one for *v* is as follows:

$$v(do(a, s)) = u \equiv \neg\exists z(a = up(z)) \wedge u = v(s) \vee \\ \exists z(a = up(z) \wedge u = v(s) + z). \tag{6.5}$$

- For the sensor device, suppose its error model is given as follows:

$$l(sonar(z), s) = u \equiv (z \geq 0 \wedge u = \mathcal{N}(z - h(s); 0, 4)) \vee (z < 0 \wedge u = 0). \tag{6.6}$$

The error model says that for nonnegative z readings, the difference between the reading and the true value is normally distributed with mean 0 (which indicates that there is no systematic bias) and variance 4.

For the remaining (physical) actions, we let

$$l(move(z), s) = 1, \quad l(up(z), s) = 1 \tag{6.7}$$

since they are assumed to be deterministic for this section.

Example 6.9 Let $\mathcal{D}$ be a basic action theory that is the union of (6.3), (6.4), (6.1), (6.5), (6.6), and (6.7). Then the following are logical entailments of $\mathcal{D}$:

1. $Bel([h = 3 \vee h = 4 \vee h = 7], S_0) = 0$.
   To see how this follows, let us begin by expanding $Bel([h = 3 \vee h = 4 \vee h = 7], S_0)$:

   $$\frac{1}{\gamma} \int_{\vec{x}} P(\vec{x}, h = 3 \vee h = 4 \vee h = 5, S_0). \tag{a}$$

   For the rest of the section, let h take its value from x_1 and v take its value from x_2. By means of (P3), there is exactly one situation for any set of values for x_1 and x_2. The P term for any such situation, however, is 0 unless $h = 3 \vee h = 4 \vee h = 5$ holds at the situation. Thus, (a) basically simplifies to:

   $$\frac{1}{\gamma} \int_{\vec{x}} \begin{cases} 0.1 \times \mathcal{N}(x_2; 0, 16) & \text{if } x_1 \in \{3, 4, 5\} \\ 0 & \text{otherwise} \end{cases} = 0.$$

   In effect, although we are integrating a function $\delta(x_1, x_2)$ over all real values, $\delta(x_1, x_2) = 0$ unless $x_1 \in \{3, 4, 7\}$.

2. $Bel(h \leq 9, S_0) = 0.7$.
   We might contrast this with the previous property in that for any given value for x_1 and x_2, the P term is 0 when $x_1 > 9$. When $x_1 \leq 9$, however, the p value for the situation is obtained from the specification given by (6.3). That is, we have:

   $$\frac{1}{\gamma} \int_{\vec{x}} \begin{cases} \text{Initial specification given by (6.3)} & \text{if } h \leq 9 \\ 0 & \text{otherwise} \end{cases}$$

   $$= \frac{1}{\gamma} \int_{\vec{x}} \begin{cases} 0.1 \times \mathcal{N}(x_2; 0, 16) & \text{if } \exists \iota.\ h(\iota) = x_1, x_1 \in [2, 12] \text{ and } h(\iota) \leq 9 \\ 0 & \text{otherwise} \end{cases}$$

   $$= \frac{1}{\gamma} \int_{\mathbb{R}} \int_{2}^{9} 0.1 \times \mathcal{N}(x_2; 0, 16)\ dx_1\ dx_2$$

   The numerator evaluates to 0.7, and the denominator to 1.

3. $Bel(h > 7v, S_0) \approx 0.6$.

Beliefs about any mathematical expression involving the random variables, even when that does not correspond to well known density functions, are entailed. To evaluate this one, for example, observe that we have

$$\frac{1}{\gamma} \int_{\vec{x}} \begin{cases} 0.1 \times \mathcal{N}(x_2; 0, 16) & \text{if } x_1 \in [2, 12] \text{ and } x_1 > 7x_2 \\ 0 & \text{otherwise} \end{cases}$$

$$= \frac{1}{\gamma} \int_2^{12} \int_{-\infty}^{x_1/7} 0.1 \times \mathcal{N}(x_2; 0, 16) dx_1.$$

4. $Bel(h = 0, do(move(4), S_0)) = 0.2$.

Here a *continuous* distribution evolves into a *mixed* distribution. This results from $Bel(h = 0, do(move(4), S_0))$ first expanding as:

$$\frac{1}{\gamma} \int_{\vec{x}} P(\vec{x}, h = 0, do(move(4), S_0)) \tag{a}$$

The P term, then, simplifies to:

$$\langle \iota. \bigwedge f(\iota) = x \wedge (h = 0)[do(move(4), \iota)] \rightarrow p(\iota, S_0) \rangle \tag{b}$$

That is, since $move(z)$ has no error component, $l(move(z), s) = 1$ for any s in accordance with $\mathcal{D}$. Therefore, $p(do(a, \iota), do(a, S_0)) = p(\iota, S_0)$. Now (b) says that for every possible value for h and v, if there is an initial situation where $h = 0$ holds after moving leftwards, then its p value is to be considered. Note that for any initial situation s where $h(s) \in [2, 4]$, we get $h(do(move(4), s)) = 0$ by (6.1). This leaves us with:

$$\frac{1}{\gamma} \int_{\vec{x}} \begin{cases} 0.1 \times \mathcal{N}(x_2; 0, 16) & \text{if } \exists \iota. h(\iota) = x_1, x_1 \in [2, 12], h(\iota) \in [2, 4] \\ 0 & \text{otherwise} \end{cases}$$

$$= \frac{1}{\gamma} \int_{\vec{x}} \begin{cases} 0.1 \times \mathcal{N}(x_2; 0, 16) & \text{if } x_1 \in [2, 4] \\ 0 & \text{otherwise} \end{cases} \tag{c}$$

We can show that $\gamma = 1$, which means (c) = 0.2. This change in beliefs is shown in Fig. 6.2.

5. $Bel(h \leq 3, do(move(4), S_0)) = 0.5$.

Bel's definition is amenable to a set of h values, where one value has a weight of 0.2, and all the other real values have a uniformly distributed density of 0.1.

6. $Bel([\exists a, s. now = do(a, s) \wedge h(s) > 1], do(move(4), S_0)) = 1$.

It is possible to refer to earlier or later situations using now as the current situation. This says that after moving, there is full belief that $(h > 1)$ held before the action.

7. $Bel(h = 4, do([move(4), move(-4)], S_0)) = 0.2$
$Bel(h = 4, do([move(-4), move(4)], S_0)) = 0$.

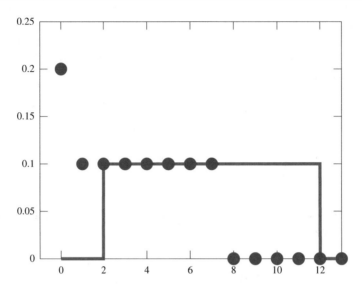

Fig. 6.2 Belief update for h after physical actions. Initial belief at S_0 (in solid red) and after a leftward move of 4 units (in blue with point markers)

The point $h = 4$ has 0 weight initially, as shown in item 1. Roughly, if the agent were to move leftwards first then many points would "collapse", as shown in item 4. The point would then obtain a h value of 0, and have a weight of 0.2. The weight is then retained on moving away by 4 units, where the point once again gets h value 4. On the other hand, if this entire phenomena were reversed then none of these features are observed because the collapsing does not occur and the entire space remains fully continuous.

8. $Bel(-1 \leq v \leq 1, do(move(4), S_0)) = Bel(-1 \leq v \leq 1, S_0) = \int_{-1}^{1} \mathcal{N}(x_2; 0, 16) dx_2$.
   Owing to Reiter's solution to the frame problem, belief in v is unaffected by a lateral motion. That is, a leftwards motion does not change v in accordance with (6.5). As per (6.3), the initial belief in $v \in [-1, 1]$ is the area between $[-1, 1]$ bounded by the specified Gaussian.

9. $Bel(v \leq 7, do(up(2.5), S_0)) = Bel(v \leq 4.5, S_0)$.
   After the action $up(2.5)$, the Gaussian distribution for v's value has its mean "shifted" by 2.5 because the density associated with $v = x_2$ initially is now associated with $v = x_2 + 2.5$. Intuitively, we have:

$$\frac{1}{\gamma} \int_{\bar{x}} \begin{cases} 0.1 \times \mathcal{N}(x_2; 0, 16) & \text{if } \exists \iota.v(\iota) = x_2 \text{ and } (v \le 7)[do(up(2.5), \iota)] \\ 0 & \text{otherwise} \end{cases}$$

$$= \frac{1}{\gamma} \int_{\bar{x}} \begin{cases} 0.1 \times \mathcal{N}(x_2; 0, 16) & \text{if } \exists \iota.v(\iota) = x_2 \text{ and } v(\iota) \le 4.5 \\ 0 & \text{otherwise} \end{cases}$$

$$= \frac{1}{\gamma} \int_{\bar{x}} \begin{cases} 0.1 \times \mathcal{N}(x_2; 0, 16) & \text{if } x_2 \le 4.5 \\ 0 & \text{otherwise} \end{cases}$$

10. $Bel(h \le 9, do(sonar(5), S_0)) \approx 0.97$.

$Bel(h \le 9, do([sonar(5), sonar(5)], S_0)) \approx 0.99$.

Compared to item 2, belief in $h \le 9$ is sharpened by obtaining a reading of 5 on the sonar, and sharpened to almost certainty on a second reading of 5. This is because the p function, according to (P2), incorporates the likelihood of each $sonar(5)$ action. More precisely, the belief term in the first entailment simplifies to:

$$\frac{1}{\gamma} \int_{\bar{x}} \langle \iota.h(\iota) = x_1 \wedge v(\iota) = x_2 \wedge h(\iota) \le 9 \to p(\iota, S_0) \times \mathcal{N}(5 - x_1; 0, 4) \rangle \qquad (a)$$

Note that we have replaced $(h \le 9)[do(sonar(5), S_0)]$ by $(h \le 9)[\iota]$ since $sonar(z)$ does not affect h. From (a), we get

$$\frac{1}{\gamma} \int_{\bar{x}} \mathcal{N}(5 - x_1; 0, 4) \times \langle \iota.h(\iota) = x_1 \wedge v(\iota) = x_2 \wedge h(\iota) \le 9 \to p(\iota, S_0) \rangle \qquad (b)$$

We know from (6.3) that those initial situations where $h \le 2$ have p values 0. Therefore, from (b), we get:

$$\frac{1}{\gamma} \int_{\bar{x}} \mathcal{N}(5 - x_1; 0, 4) \times \begin{cases} 0.1 \times \mathcal{N}(x_2; 0, 16) & \text{if } x_1 \in [2, 9] \\ 0 & \text{otherwise} \end{cases}$$

$$= \frac{1}{\gamma} \int_{\mathbb{R}} \int_2^9 \mathcal{N}(5 - x_1; 0, 4) \times 0.1 \times \mathcal{N}(x_2; 0, 16) dx_1 dx_2$$

After a second reading of 5 from the sonar, the expansion for belief is analogous, except that the function to be integrated gets multiplied by a second $\mathcal{N}(5 - x_1; 0, 4)$ term. It is then not hard to see that belief sharpens significantly with this multiplicand. The agent's changing densities are shown in Fig. 6.3.

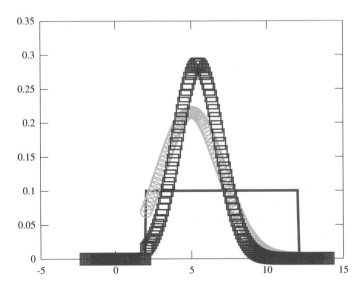

Fig. 6.3 Belief change for h at S_0 (in solid red), after sensing 5 (in green with circular markers), and after sensing 5 twice (in blue with rectangle markers)

6.5   Noisy Acting

In the presentation so far, we assumed physical actions to be deterministic. By this we mean that when a physical action a occurs, it is clear to us (as modelers) but also the agent how the world has changed on a. Of course, in realistic domains, especially robotic applications, this is *not* the usual case. In this section, in a domain that has continuous fluents, we show how our current account of belief can be extended to reason with sensors as well as effectors that are noisy.

In line with the rest of this work, effector noise is given a quantitative account. Let us first reflect on what is expected with noisy acting. When an agent senses, as in the case of $sonar(z)$, the argument for this action is not chosen by the agent. That is, the world decides what z should be, and based on this reading of z, the agent comes to certain conclusions about its own state. The noise factor, then, simply addresses the phenomena that the number z returned may differ from the true value of whatever fluent the sensor is measuring.

Noisy acting diverges from that picture in the following sense. The agent *intends* to do action a, but what actually occurs is a' that is possibly different from a. For example, the agent may want to move 3 units, but, unbeknownst to the agent, it may move by 3.042 units. The agent, of course, does not observe this outcome. Nevertheless, provided the agent has an account of its effector's inaccuracies, it is reasonable for the agent to believe that it is in fact closer to the wall, even if it may not be able to precisely tell by how much. Intuitively, the result of a nondeterministic action is that any number of successor situations might be obtained, which are all indistinguishable in the agent's perspective (until sensing is performed). Depending on the likelihoods of the action's potential outcomes, some of these successor situations are considered more probable than others. The agent's belief about what holds then must incorporate these relative likelihoods. So, in our view, nondeterminism is really an *epistemic* notion.

6.5.1   Noisy Action Types

Perhaps the simplest extension to make all this precise is to assume that deterministic actions such as $move(x)$ now have companion action types $move(x, y)$. The intuition is that x represents the *nominal* value, which is the number of units that the agent intends to move, while y represents the actual distance moved. The actual value of y in any ground action, of course, is not observable for the agent. This simple idea will need three adjustments to our account:

1. successor state axioms need to be built using these new action types;
2. the formalism must allow the modeler to formalize that certain outcomes are more likely than others, that is, noisy actions may be associated with a probabilistic account of the various outcomes; and
3. the notion of belief must incorporate the nominal value, the range of possible outcomes and their likelihoods.

First, we address successor state axioms. These are now specified as usual, but using the second argument, which is the actual outcome, rather than the nominal value, which is ignored. For example, for the fluent h, instead of (6.1), we will now have:

$$h(do(a, s)) = u \equiv \exists x, y[a = move(x, y) \land u = \max(0, h - y)] \lor \\ \neg \exists x, y[a = move(x, y)] \land u = h(s). \tag{6.8}$$

The reason for this modification is obvious. If y is the actual outcome then the fluent change should be contingent on this value rather than what was intended. It is important to note that no adjustment to the existing (Reiter's) solution to the frame problem is necessary.

6.5.2   The GOLOG Approach

The foremost issue now is to use the above idea to allow for more than one possible successor situation. Clearly, we do not want the agent to control the actual outcome in general. One approach is to think of picking the second argument as a *nondeterministic* GOLOG program. Recall from Sect. 3.2.6, GOLOG is an agent programming proposal where one is allowed to formulate complex actions that denote sequential and nondeterministic executions of actions, among others, and is essentially a basic action theory. Given the action $move(x, y)$, for example, the GOLOG program MOVE(X) might stand for the abbreviation $\pi y.\ move(x, y)$, which corresponds to a ground action $move(x, n)$ where n is chosen nondeterministically. For our purposes, we would then imagine that the agent executes GOLOG programs.

There are some advantages to this approach: namely, we only have to look at the logical entailments, including ones mentioning *Bel*, of such GOLOG programs. Since traces of these programs account for many potential outcomes, *Bel* does the right thing and accommodates all of these when considering knowledge. But the disadvantage is that the resulting formal specification turns out be unnecessarily complex, at least as far as projection is concerned.

For projection tasks, we show that we can settle on a simpler alternative, one that does not appeal to GOLOG. We still assume that the world is deterministic, where the result of doing a ground action leads to a distinct successor. The idea is that when a noisy action is performed, the various outcomes of the action as well as the potential successor situations that are obtained w.r.t. these are treated at the level of belief.

6.5.3 Alternate Action Axioms

Our approach is based on the introduction of a distinguished predicate *Alt*. The idea is this: if $Alt(a, a', \vec{z})$ holds for ground action $a = A(\vec{c})$ then we understand this to mean that the agent believes that any instance of $a' = A(\vec{z})$ might have been executed instead of a. Here, $\vec{z}$ denotes the range of the arguments for potential outcomes.

To see how that gets used with the new action types such as $move(x, y)$, consider the ground action $move(3, 3.1)$. So, the agent intends to move by 3 units but what has actually occurred is a move by 3.1 units. Since the agent does not observe the latter argument, from its perspective, what occurred could have been a move by 2.9 units, but also perhaps (although less likely) a move by 9 units. Thus, the ground actions $move(3, 2.9)$ and $move(3, 9)$ are *Alt*-related to $move(3, 3.1)$. (The likelihoods for these may vary, of course.) In logical terms, we might have have an axiom of the following form in the background theory:

$$Alt(move(x, y), a', z) \equiv a' = move(x, z). \tag{6.9}$$

This to be read as saying that $move(3, z)$ for every $z \in \mathbb{R}$ are alternatives to $move(3, 3.1)$. If we required that z is only a certain range from 3, for example, we might have:

$$Alt(move(x, y), a', z) \equiv a' = move(x, z) \wedge |x - z| \le c$$

where c bounds the magnitude of the maximal possible error. On the other hand, for actions such as $sonar(z)$, which do not have any alternatives, we simply write:

$$Alt(sonar(x), a', z) \equiv a' = sonar(z) \wedge z = x. \tag{6.10}$$

This says that $sonar(x)$ is only *Alt*-related to itself.

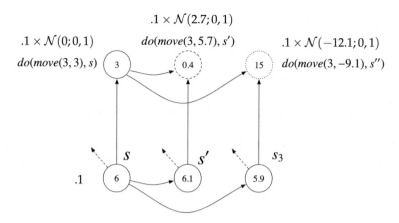

Fig. 6.4 Situations with accessibility relations after noisy acting. The numbers inside the circles denote the h values at these situations. Dotted circles denote a lower density in relation to their epistemic alternatives

With this simple technical device, one can now additionally constrain the likelihood of various outcomes using l. For example:

$$l(move(x, y), s) = u \equiv u = \mathcal{N}(y - x; \mu, \sigma^2) \tag{6.11}$$

says that the difference between nominal value and the actual value is normally distributed with mean μ and variance σ^2. This essentially corresponds to the standard additive Gaussian noise model.

To see an example of how, say, (6.9) and (6.11) work together with the successor state axiom (P2) for p, consider three situations s, s' and s'' associated with the same density, as shown in Fig. 6.4. Suppose their h values are 6, 6.1 and 5.9 respectively. After attempting to move 3 units, the action $move(3, z)$ for any $z \in \mathbb{R}$ may have occurred. So, for each of the three situations, we explore successors from different values for z. Assume the motion effector is defined by a mean $\mu = 0$ and variance $\sigma^2 = 1$. Then, the p-value of the situation $do(move(3, 5.7), s')$, for example, is obtained from the p-value for s' multiplied by the likelihood of $move(3, 5.7)$, which is $\mathcal{N}(5.7 - 3; 0, 1) = \mathcal{N}(2.7; 0, 1)$. Thus, the successor situation $do(move(3, 5.7), s')$ is much less likely than the successor situation $do(move(3, 3), s)$, as should be the case.

In general, we define *alternate actions* axioms that are to be a part of the basic action theory henceforth:

Definition 6.10 Let $A(\vec{x}, \vec{y})$ be any action. *Alternate actions axioms* are sentences of the form:

$$Alt(A(\vec{x}, \vec{y}), a', \vec{z}) \equiv a' = A(\vec{x}, \vec{z}) \wedge \psi(\vec{x}, \vec{y}, \vec{z})$$

where ψ is a formula that characterizes the relationship between the nominal and true values.

The one limitation with this definition is that only actions of the same *type*, i.e., built from the same function symbol, are alternatives to each other. This does not allow, for example, situations where the agent intends a physical move, but instead unlocks the door. Nevertheless, this definition is not unreasonable because noisy actions in robotic applications typically involve additive noise. Moreover, this limitation only assists us in arriving at a simple and familiar definition for belief. A more involved definition would allow for other variants.

6.5.4   A Definition for Belief

We have thus far successfully augmented successor state axioms and extended the formalism for modeling noisy actions. The final question, then, is how can the outcomes of a noisy action, and their likelihoods, be accounted for? Indeed, a formula might not only be true as a result of the actions intended, but also as a result of those that were not.

Consider the simple case of deterministic actions, where the density associated with s is simply transferred to $do(a, s)$. In contrast, if a and a' are Alt-related, then the result of doing a at s would lead to successor situations $do(a, s)$ and $do(a', s)$. Moreover, unlike noisy sensors, a and a' may affect fluent values in different ways, which is certainly the case with $move(3, 3.074)$ and $move(3, 3)$ on the fluent h. Thus, the idea then is that when reasoning about the agent's beliefs about ϕ, one would need to integrate over the densities of all those potential successors where ϕ would hold.

To make this precise, let us first consider the result of doing a single action a at S_0. The *degree of belief* in ϕ after doing a is now an abbreviation for:

$$Bel(\phi, do(a, S_0)) \doteq \frac{1}{\gamma} \int_{\vec{x}} \int_z P(\vec{x}, z, \phi, do(a, S_0))$$

where

$$P(\vec{x}, z, \phi, do(a, S_0)) \doteq \langle \iota, b . \bigwedge f_i(\iota) = x_i \wedge Alt(a, b, z) \wedge \phi[do(b, \iota)] \rightarrow p(do(b, \iota), do(a, S_0)) \rangle.$$

As before, the i ranges over the indices of the fluents, that is, $\{1, \ldots, n\}$. The intuition is this. Recall that by integrating over $\vec{x}$, all possible initial situations are considered by $f_i(\iota) = x_i$. Analogously, by integrating over z, all possible action outcomes are considered by $Alt(a, b, z)$. Supposing $a = A(x, y)$, for each outcome $b = A(x, z),$[3] we test whether ϕ holds at the resulting situation $do(b, \iota)$ as before, and use its p-value. Here, this p-value is

---

[3] For ease of presentation, we assume that the nominal and the actual arguments involve a single variable.

given by $p(do(b, \iota), do(a, S_0))$, where the first argument is the successor of interest $do(b, \iota)$ and the second is the real world $do(a, S_0)$.

The generalization, then, for a sequence of actions is as follows:

Definition 6.11 (*Degrees of belief (continuous noisy effectors and sensors.)*) Suppose ϕ is any situation-suppressed formula. The *degree of belief* in ϕ at s, written $Bel(\phi, s)$, is an abbreviation:

$$Bel(\phi, s) \doteq \frac{1}{\gamma} \int_{\vec{x}} \int_{\vec{z}} P(\vec{x}, \vec{z}, \phi, s)$$

where, if $s = do([a_1, \ldots, a_k], S_0)$, then

$$P(\vec{x}, \vec{z}, \phi, s) \doteq$$
$$\langle \iota, b_1, \ldots, b_k. \bigwedge f_i(\iota) = x_i \wedge \bigwedge Alt(a_j, b_j, z_j) \wedge$$
$$\phi[do([b_1, \ldots, b_k], \iota)] \to p(do([b_1, \ldots, b_k], \iota), do([a_1, \ldots, a_k], S_0))\rangle.$$

Here, i ranges over the fluent indices $\{1, \ldots, n\}$ as before, and j ranges over the indices of the ground actions $\{1, \ldots, k\}$. That is, given any sequence, for all possible $\vec{z}$ values, we consider alternate sequences of ground action terms and integrate the densities of successor situations that satisfy ϕ, using the appropriate p-value.[4]

6.5.5 Axiomatization: The Robot with Noisy Effectors

Let us now build a simple example with noisy actions. Consider the robot scenario in Fig. 3.2. Suppose the basic action theory $\mathcal{D}$ includes the foundational axioms, and the following components.

The initial theory $\mathcal{D}_0$ includes the following p specification:

$$p(\iota, S_0) = \begin{cases} 0.5 & \text{if } 10 \leq h(\iota) \leq 12 \\ 0 & \text{otherwise} \end{cases} \quad (6.12)$$

For simplicity, let h be the only fluent in the domain, and assume that actions are always executable. The successor state axiom for the fluent h is (6.8). For p, it is the usual one, viz. (P2).

We imagine two actions in this domain, one of which is the noisy move $move(x, y)$ and a sonar sensing action $sonar(z)$. For the alternate actions axioms, let us use (6.9) and (6.10).

Finally, we specify the likelihood axioms. Let the sonar's error profile be:

$$l(sonar(z), s) = u \equiv (z \geq 0 \wedge u = \mathcal{N}(z - h(s); 0, 0.25)) \vee (z < 0 \wedge u = 0). \quad (6.13)$$

---

[4] In later chapters, for readability purposes we abuse notation and sometimes lump both types of variables as a single vector by concatenating them to $\vec{x} \cdot \vec{z}$. Consequently, we have a density term with only three arguments: $P(\vec{x} \cdot \vec{z}, \phi, s)$, and only need to use a single integration symbol.

Readers may note that this sonar is more accurate than the one characterized by (6.6), as it has a smaller variance. Regarding the likelihood axiom for $move(x, y)$, let that be:

$$l(move(x, y), s) = u \equiv u = \mathcal{N}(y - x; 0, 1).$$
(6.14)

This completes the specification of $\mathcal{D}$.

Example 6.12 The following are entailments of $\mathcal{D}$ above:

1. $Bel(h \geq 11, do(move(-2, -2.01), S_0)) \approx 0.95$
   We first observe that for calculating the degrees of belief, we have to consider all those successors of initial situations w.r.t. $move(-2, z)$ for every z, where ϕ holds. By (P2), the p value for such situations is the initial p value times the likelihood of $move(-2, z)$, which is $\mathcal{N}(z + 2; 0, 1)$ by (6.14). Therefore, we get

$$\frac{1}{\gamma} \int_x \int_z \begin{cases} 0.5 \times \mathcal{N}(z + 2; 0, 1) & \text{if } \dots \\ 0 & \text{otherwise} \end{cases}$$

   where the ellipsis is for $\exists \iota. x \in [10, 12], h(\iota) = x$ and $(h \geq 11)[do(move(-2, z), \iota)]$. We get:

$$\frac{1}{\gamma} \int_x \int_z \begin{cases} 0.5 \times \mathcal{N}(z + 2; 0, 1) & \text{if } \exists \iota. x \in [10, 12], h(\iota) = x \text{ and } h(\iota) - z \geq 11 \\ 0 & \text{otherwise} \end{cases}.$$

$$= \frac{1}{\gamma} \int_{-\infty}^{\infty} \int_{10}^{12} \begin{cases} 0.5 \times \mathcal{N}(z + 2; 0, 1) & \text{if } x \in [10, 12] \text{ and } x - z \geq 11 \\ 0 & \text{otherwise} \end{cases}$$

   It is not hard to see that had the action been deterministic, the degree of belief in $h \geq 11$ after moving away by 2 units should have been precisely 1. In Fig. 6.5, we see the effect of this move, where the range of h values with non-zero densities extends considerably more than 2 units.

2. $Bel(h \geq 10, do([move(-2, -2.01), move(2, 2.9)], S_0)) \approx 0.74$
   The argument proceeds in a manner identical to the previous demonstration. The density function is further multiplied by a factor of $\mathcal{N}(u - 2; 0, 1)$, from (6.14) and (P2). More precisely, we have

$$\frac{1}{\gamma} \int_x \int_z \int_u \begin{cases} 0.5 \times \mathcal{N}(z + 2; 0, 1) \times \mathcal{N}(u - 2; 0, 1) & \text{if } \dots \\ 0 & \text{otherwise} \end{cases}$$

   where the ellipsis is for $\exists \iota. x \in [10, 12], h(\iota) = x$ and $h(\iota) - z - u \geq 10$. We get:

$$\frac{1}{\gamma} \int_z \int_u \int_{10}^{12} \begin{cases} 0.5 \times \mathcal{N}(z + 2; 0, 1) \times \mathcal{N}(u - 2; 0, 1) & \text{if } x \in [10, 12] \text{ and } x - z - u \geq 10 \\ 0 & \text{otherwise} \end{cases}$$

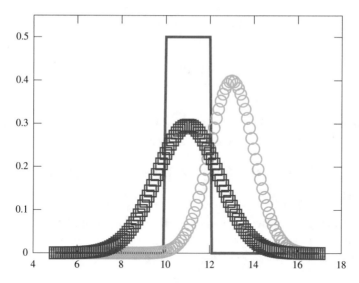

Fig. 6.5 Belief change for h at S_0 (in solid red), a noisy move away from the wall (in green with circular markers), and after a second noisy move towards the wall (in blue with rectangle markers)

If the action were deterministic, yet again the degree of belief about $h \geq 10$ would be 1 after the intended actions. That is, the robot moved away by 2 units and then moved towards the wall by another 2 units, which means that h's current value should have been precisely what the initial value was.

See Fig. 6.5 for the resulting density change. Intuitively, the resulting density changes as effectuated by the moves degrades the agent's confidence considerably. In Fig. 6.5, for example, we see that in contrast to a single noisy move, the range of h values considered possible has extended further, leading to a wide curve.

3. $Bel(h \geq 11, do([move(-2, -2.01), sonar(11.5)], S_0) \approx 0.94$

   This demonstrates the result of a sensing action after a noisy move. Using arguments analogous to those in the previous item, it is not hard to see that we have:

$$\frac{1}{\gamma} \int_x \int_z \begin{cases} 0.5 \times \mathcal{N}(z + 2; 0, 1) \times \mathcal{N}(x - z - 11.5; 0, 0.25) & \text{if } \dots \\ 0 & \text{otherwise} \end{cases}$$

   where the ellipsis is for $\exists \iota. \, x \in [10, 12]$, $h(\iota) = x$ and $h(\iota) - z \geq 11$.

4. $Bel(h \geq 11, do([move(-2, -2.01), sonar(11.5), sonar(12.6)], S_0) \approx 0.99$

   In this case, two successive readings around 12 strengthens the agent's belief about $h \geq 11$. The density function is multiplied by $\mathcal{N}(x - z - 12.6; 0, 0.25)$ because of (P2) and (6.6) as follows:

$$\frac{1}{\gamma} \int_x \int_z \begin{cases} \delta \times \mathcal{N}(x - z - 11.5; 0, 0.25) \times \mathcal{N}(x - z - 12.6; 0, 0.25) & \text{if } \delta' \\ 0 & \text{otherwise} \end{cases}$$

$$= \frac{1}{\gamma} \int_z \int_x \begin{cases} \delta \times \mathcal{N}(x - z - 11.5; 0, 0.25) \times \mathcal{N}(x - z - 12.6; 0, 0.25) & \text{if } \delta'' \\ 0 & \text{otherwise} \end{cases}$$

where $\delta = 0.5 \times \mathcal{N}(z + 2; 0, 1)$, δ' is $\exists \iota. x \in [10, 12]$, $h(\iota) = x$, $h(\iota) - z \geq 11$, and δ'' is $x \in [10, 12]$ and $x - z \geq 11$. The normalization constant γ is

$$\int_z \int_x \begin{cases} \delta \times \mathcal{N}(x - z - 11.5; 0, 0.25) \times \mathcal{N}(x - z - 12.6; 0, 0.25) & \text{if } x \in [10, 12] \\ 0 & \text{otherwise} \end{cases}$$

In Fig. 6.6, the agent's increasing confidence is shown as a result of these sensing actions. Note that even though the sensors are noisy, the agent's belief about h's true value sharpens because the sensor is a fairly accurate one.

In the following chapters, we expand on these examples to axiomatize the well-understood problem of localizing a robot. We will then turn to computational schemes for eliminating actions from a projection query: that is, the equivalent of regression and progression for *Bel*.

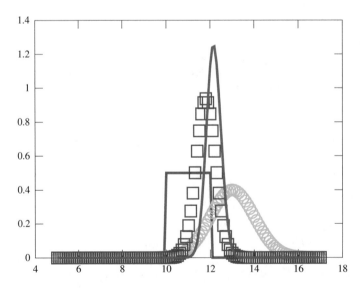

Fig. 6.6 Belief change for h at S_0 (in solid red), after a noisy move (in green with circular markers), after sensing once (in blue with rectangle markers), and after sensing twice (in solid magenta)

6.6   Notes

This chapter generalized the existing scheme of BHL [8] to handle both discrete and continuous probability distributions while retaining all the advantages; see [28] for the full account, including proofs and generalizations to mixtures of discrete and continuous distributions.[5]

Related efforts were discussed in the previous chapter, and so we will not reiterate them here. At the outset, note that there are very few first-order accounts to allow continuous probability distributions, and none that accommodate a general model of knowledge and actions.[6]

In contrast, albeit in a non-logical setting, the use of continuous event and observation variables are commonplace in robotics and machine learning applications; see, for example, [173, 219]. The book on probabilistic robotics [219] is also an excellent source on notions such as additive noise, localization and other concerns with noisy sensing.

As discussed in [28], the use of a distinguished predicate for capturing noisy actions is inspired by [62]. Making physical actions distinct from sensing ones in the situation calculus is primarily for convenience, especially in matters of formalization, following [202]. The axiomatization for multiple initial situations is from [153].

Our formalization rested on assuming finitely many fluent symbols. Thus, we needed to enforce a representational limitation that is fairly serious from a relational viewpoint.[7] Lifting this limitation would make the proposal undoubtedly richer. Nonetheless, for subsequent chapters, we will persist in using this continuous variant for our exercises, from axiomatizing localisation to addressing projection to designing a programmatic model. We do this for two reasons. First, it becomes vastly simpler to present results such as regression by studying how noise and likelihood affect random variables. Readers more familiar with probability theory in particular would see treatments involving individual random variables as natural too. Conveniently, our simple robot example does not demand much more. Overall, quantifiers in the initial theory would needlessly complicate the presentation for understanding changing degrees of belief. All these results could be made to work with the full expressiveness of the situation calculus with discrete probabilities (that is, the BHL scheme) with some effort.

The second reason to focus on continuous models is because the vast majority of logical formalisms admitting probabilities only focus on discrete probabilities. In that regard, a continuous model is a refreshing change. Indeed, studying how densities interoperate with a

---

[5] We note that the definition of *Bel* can be generalized further, to allow for combinations of discrete and continuous sensors and effectors. The main idea is to simply allow the range of some fluents and action outcomes to be taken from finite sets, and appropriately interleave integrals and sums. This is relatively straightforward, but somewhat tedious. See [28] for details. For most purposes, however, we can continue to use Definition 6.11 and interpret the integral as a sum when appropriate.

[6] For one of earliest accounts on reasoning about probability and knowledge, albeit in a propositional setting, see [72].

[7] It remains to be seen whether ideas from probability theory on high dimensions [31, 53] and infinite-dimensional probabilistic graphical models [205] can be leveraged for our purposes.

logical theory of action could serve as a guide for other such logical models—in particular, when discussing regression, we contrast how regressing with discrete probabilities is then generalized to continuous models. Likewise, when discussing the implementation semantics for a programmatic model with degrees of belief, we see the necessary role that sampling has to play in the presence of countably infinite worlds. All such discussions become possible by committing to the continuous extension.

Localization

7

Another flaw in the human character is that everybody wants to build and nobody wants to do maintenance.

– ***Kurt Vonnegut**, Hocus Pocus*

Cognitive robotics is the high-level control paradigm that attempts to apply knowledge representation (KR) technologies to the reasoning problems faced by an autonomous agent, such as a robot, in an incompletely known dynamic world. Although a tight pairing of sensor data and high-level control is generally what is desired, typical sensor data is best treated *probabilistically* while most knowledge change accounts (especially in the knowledge representation community) are *categorical*, that is, they do not represent and reason about changing degrees of belief. This has led to a major criticism that these logical formalisms are not realistic for applications involving actual physical robots. Indeed, a domain designer is now left with the rather unfortunate task of modeling probabilistic sensors in terms of non-probabilistic ones, an extreme being noise-free sensors, which would lead to an inaccurate model. More drastically, designers may completely abandon the inner workings of sensors to black-box probabilistic tools, in which case their outputs would need to be interpreted in some qualitative fashion and so is not straightforward. Regardless of application domains where such a move might be appropriate, for computational reasons or otherwise, both of these limitations are very serious since they challenge the underlying theory as a genuine characterization of the agent. Other major concerns include: (a) the loss of granularity, as it is not clear at the outset which aspect of the sensor data is being approximated and by how much, and (b) the domain designer is at the mercy of her intuition to imagine the various ways sensors might get used.

The formalism in this book and a few others do rise up to the challenge. How can a formalism like ours address fundamental capabilities of modern robotics systems, such as *localization*, where a robot uses its noisy sensors to situate itself in a dynamic world? Indeed, it is one thing to model low-level sensors, which addresses the granularity question,

© The Author(s), under exclusive license to Springer Nature Switzerland AG 2023
V. Belle, *Toward Robots That Reason: Logic, Probability & Causal Laws*,
Synthesis Lectures on Artificial Intelligence and Machine Learning,
https://doi.org/10.1007/978-3-031-21003-7_7

but another to capture a probabilistic robotics application. In fact, early breakthrough in probabilistic robotics was precisely in getting robots localized, and although the application is not difficult and standard fare in today's robots, it is surprisingly not addressed by any other logical framework.

In this chapter, we revisit the localization problem and axiomatize it in the situation calculus. This serves at least two purposes. On the one hand, it is an opportunity for the knowledge representation community to understand what is needed to address localization in a native manner. On the other, it showcases the flexibility afforded by such axiomatizations to the traditional robotics community. The purpose of this chapter is not, however, to suggest that such an axiomatization is how we should compute beliefs in practice. Rather it is only showing that the language is representational rich enough, that it is possible to develop a fully logical account of the robot's mental state, and that localization is not an isolated adventure from modeling degrees of belief.

7.1   Axiomatization

One of the significant features about our proposal is that robot localization, among other capabilities, follows logically from a basic action theory. No new foundational axioms are necessary. In fact, localization is a certain degree of belief regarding position and orientation, and so by reasoning about belief change in terms of *projection*, the robot would get localized. On the one hand, this is perhaps expected as many state estimation techniques in robotics are based on Bayesian conditioning, but on the other, we are demonstrating this capability in a very rich first-order framework.

In this section, we develop a simple example, and a basic action theory corresponding to this example. Localization will then be demonstrated in terms of logical entailments of the action theory. We think many of the features of our example are suggestive of how one would approach more complex domains. In the main, the example involves the following steps:

- a characterization of the environment (walls, doors, etc.);
- a characterization of the uncertainty of the robot about this environment (its position and orientation); and
- a characterization of the robot's actions and sensors, and how they depend on and affect the environment.

Our example imagines a robot in a two-dimensional grid, equipped with a moving action and distance sensor, and facing two parallel walls as in Fig. 7.1. The basic action theory $\mathcal{D}$ developed for the above items of this domain will be built using three fluents h (horizontal position), v (vertical position) and θ (orientation) that will determine the pose of the robot, a single rigid predicate *Solid* used to axiomatize the environment, two action types $move(z, w)$ and $rotate(z)$ that determine how the robot moves and how these affect the fluents using

Fig. 7.1 Two walls and a robot

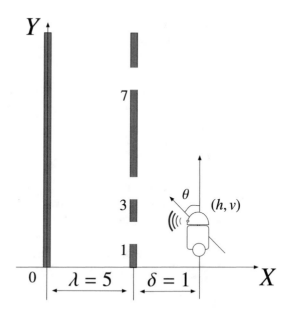

successor state axioms, a single sensing action *sonar(z)*, and convenient abbreviations that expand into formulas involving the aforementioned logical symbols. Of course, we assume $\mathcal{D}$ to also mention *Poss*, *l* and *p*, which are distinguished symbols. We reiterate that we will not need any machinery beyond the standard situation calculus from the previous chapters.

7.1.1 Environment

The very first item on the agenda is the notion of a *map*, which for our purpose will simply mean an axiomatic formulation of the physical space. In our example, suppose that the two parallel walls are 10 units long. The one on the *extreme* left of the robot, which we refer to as WALLCLOSE in the sequel, is without any doors, while the one that is adjacent to the robot, referred to as WALLFAR, has 3 open doors. The doors extend for one unit each. As can be seen in Fig. 7.1, we are imagining a coordinate system that has WALLCLOSE on the Y-axis, and puts the bottom edge of WALLCLOSE at the origin.

For our purposes, we develop a simple axiomatization to describe this physical space. We think of the walls in terms of continuous solid segments, that is, WALLCLOSE is considered to be a single chunk, while WALLFAR is thought of as 4 components. We will ignore the thickness of walls for simplicity. In precise terms, let *Solid(x, y, d)* indicate that beginning at the coordinate (x, y), one finds a solid structure of length d extending from (x, y) to $(x, y + d)$. Of course, we are using a rigid predicate because walls are stationary; for dynamic objects, such as the robot, fluents will be used. With this idea, we could characterize (say)

Table 7.1 A basic action theory for the domain

1.	$\{Solid(0, 0, 10), Solid(5, 0, 1), Solid(5, 2, 1), Solid(5, 4, 3), Solid(5, 8, 2)\}$.
2.	$Poss(a) \equiv true$.
3.	$h(do(a, s)) = u \equiv$
	$\quad \neg \exists z, w(a = move(z, w)) \wedge u = h(s) \ \vee$
	$\quad \exists z, w(a = move(z, w) \wedge u = \max(\delta(s), h(s) - z \cdot \cos(w)))$.
4.	$v(do(a, s)) = u \equiv$
	$\quad \neg \exists z, w(a = move(z, w)) \wedge u = v(s) \vee$
	$\quad \exists z, w(a = move(z, w) \wedge u = v(s) + z \cdot \sin(w))$.
5.	$\theta(do(a, s)) = u \equiv$
	$\quad \neg \exists z(a = rotate(z) \wedge u = \theta(s)) \vee$
	$\quad \exists z(a = rotate(z) \wedge u = (((\theta(s) + z) \bmod 360) - 180))$.
6.	$\{l(move(z, w), s) = 1, l(rotate(z), s) = 1\}$.
7.	$l(sonar(z), s) = u \equiv$
	$\quad Blocked(s) \wedge u = \mathcal{N}(\delta/\cos(\theta) - z; 0, 1)[s] \vee$
	$\quad \neg Blocked(s) \wedge u = \mathcal{N}((\delta + \lambda)/\cos(\theta) - z; 0, 1)[s]$.

WALLCLOSE by including $Solid(0, 0, 10)$ in $\mathcal{D}_0$. For both walls, then, $\mathcal{D}_0$ is assumed to include the formulas (1) from Table 7.1.

It should be clear that one may easily extract various directional and spatial relationships between such objects as appropriate. For example, although entirely obvious here, to calculate the distance between the walls, one may define an abbreviation λ as follows:

$$\lambda = u \doteq \exists x, y, d, x', y', d'. \ Solid(x, y, d) \wedge Solid(x', y', d') \wedge$$
$$x \neq x' \wedge u = |x - x'|.$$

7.1.2   Robot: Physical Actions

Here, we characterize the robot's position, and how that is affected using physical actions.

The pose of the robot is given by three fluents: h, v and θ, where h is the horizontal position, v is the vertical position and so (h, v) is the robot's location, and θ is the orientation. We let θ range from -180 to 180 (degrees), with $\theta = 0$ indicating that the robot is perpendicular to WALLCLOSE and directed towards it, and $\theta = 90$ indicating that the robot is perpendicular to the X-axis and directed towards the positive half of the Y-axis.

We imagine two physical action types at the robot's disposal, $move(z, w)$ and $rotate(z)$. We are thinking that the robot is capable of moving z units along the orientation w (degrees) w.r.t. its angular frame. That is, for $w = 0$, the robot would move z units towards WALLCLOSE, and for $w = 90$, the robot would move z units along the positive Y-axis, i.e., parallel to WALLCLOSE. The robot can also orient itself in-place, using $rotate(z)$. For these actions, one also needs to specify their preconditions, and their likelihood axioms. In this work, we make two simplifying assumptions. First, we assume these and all other actions in domain (including the sensing action to be discussed shortly) are always executable, given by (2). Second, we assume the physical actions are noise-free. (The sensor will be noisy, however, as we shall see.) Thus, likelihood axioms, which are used to specify probabilistic nondeterminism, will be given by (6) for these actions. Essentially, (6) says that the likelihood is 1 for these actions.

The values of fluents may change after actions. The sentence (P2) already specifies how p behaves in successor situations. We now do the same for h, v and θ. Since $move(z)$ and $rotate(z)$ are the only physical actions, the successor state axioms for h, v and θ will only mention these actions. They are given as (3), (4) and (5) respectively. Let us consider them in order.

In the case of h, we would like $move(z, 0)$ to bring the robot z units towards the wall on its left, but that motion should stop if the robot hits the wall. For this, it is perhaps easiest to first infer the distance between the robot and the closest wall on its left. This can be done as follows. For an arbitrary coordinate (x^*, y^*), we define an abbreviation for the nearest wall on its left:

$$NearestLeft(x^*, y^*) = d \doteq \exists x, y, d.\ Solid(x, y, d) \wedge y^* \in [y, y+d] \wedge$$
$$\neg \exists x', y', d'.\ Solid(x', y', d') \wedge y^* \in [y', y'+d'] \wedge$$
$$(x^* - x') < (x^* - x) \wedge\ d = (x^* - x).$$

We use $u \in [v, w]$ to mean $u \geq v \wedge u \leq w$, as usual. To now extract the distance between the *robot* and the nearest wall on its left, simply define an abbreviation δ as follows:

$$\delta(s) = u \doteq u = NearestLeft(h(s), v(s)).$$

This now allows us to dissect (7.1.1). It says that $move(z, w)$ is the only action affecting h, thereby incorporating Reiter's monotonic solution to the frame problem, and it decrements h by $z \cos(w)$ units but stops if the robot hits the nearest wall on its left. Note that, then, the value of h will become δ. For example, if $\theta = 0$, then the new value of h is simply decremented by z, and if $\theta = 180$, which would mean the robot is facing away from WALLFAR then h would be incremented by z (since $\cos(180)$ is -1.)

For the fluent v, the treatment is analogous, as shown in (4). That is, the action $move(z, w)$ would increment v by $z \cdot \sin(\theta)$. For example, if $z = 90$, then the move action would simply increment v since the motion would be along the Y-axis in an incremental fashion. Naturally, if one were to give a negative argument, say -3, to $move$, then the robot would move from (h, v) to $(h, v - 3)$.

Finally, θ is manipulated using $rotate(z)$ in an incremental manner while keeping its range in $[-180, 180]$ in (5).

7.1.3 Robot: Sensors

The robot is assumed to have a sonar unit on its frontal surface, that is, along θ. We take this sensor to be noisy. What this means is that if the robot is facing WALLFAR, then a reading z from the sensor may *differ* from δ, but perhaps in some reasonable way. Most sensors have additive Gaussian noise, which is to say the likelihood of z is obtained from a normal curve whose mean is δ.

The complication here is that there are two walls and depending on the robot's pose, the sensor might be measuring either δ or $\lambda + \delta$. For example, if $h \in [0, 1]$ and $\theta = 0$, we understand that the sonar's signals would likely be centered around δ. However, if $v < 1$ but the robot's orientation is such that the sonar's signals advance through the gap at $[1, 2]$, then the robot's sonar unit would suggest values closer to $\delta + \lambda$ rather than δ alone. To provide a satisfactory l-axiom for the sensor, let us first introduce an abbreviation for what it means for a sensor's signals to stop at WALLFAR:

$$Blocked(s) \doteq \exists x, y, d. \ Solid(x, y, d) \wedge h(s) = x + \delta(s) \wedge$$
$$(v + \delta \cdot \tan(\theta))[s] \in [y, y + d].$$

To make sense of this in (converse) terms of when signals would reach WALLCLOSE, note that if $v < 1$ and yet $v + \tan(\theta) \in [1, 2]$, then the signal advances through the gap. Analogously, if $\theta < 0$ and $v > 2$ and yet $v + \tan(\theta) \in [1, 2]$, then the signal advances through as well. This then allows us to define an l axiom for the sonar in (7). Intuitively, when *Blocked* holds at situation s, we assume the sonar's reading to have additive Gaussian noise (with unit variance) centered around δ, but when the sonar's signals can reach WALLCLOSE, we assume its reading to have additive Gaussian noise (with unit variance) centered around $\delta + \lambda$. (The $\mathcal{N}$ term is an abbreviation for the mathematical formula defining a Gaussian density.)

7.1.4 Initial Constraints

The final step is to decide on a p specification for the domain. Recall that the p fluent is used to formalize the (probabilistic) uncertainty that the robot has about the domain. This perhaps accounts for a major difference between the work here and almost all probabilistic formalisms. For us, in a sense, p is just another fluent function, allowing the domain modeler to provide incomplete and partial specifications. But since our objective in this chapter will be to show, in the least, that robot localization behaves as it does in standard probabilistic formalisms, we discuss two examples with fully known joint distributions in the next section. There are other possibilities still, a discussion of which we defer to later.

7.2   Properties

Before looking at the two examples, let us briefly reflect on what is expected. A reasonable belief change mechanism would support the following: .

- Suppose the agent believes v to be uniformly distributed on the interval [0,10]. If the robot then uses its sonar and senses a value close to $\lambda + \delta$ say 5.9, it should come to believe that it is located at a door, which would deflate its beliefs about every point not in $[1, 2] \cup [3, 4] \cup [7, 8]$ (i.e., open gaps in WALLFAR.).
- Suppose the robot moves 2 units away from the X axis and then uses its sonar obtaining a reading of 5.8. It should then believe, rather confidently, that it must be in [3, 4] since that is the only trajectory that supports a door initially and a second door after 2 units.

We now confirm these intuitions below, which essentially amount to the robot situating itself.

7.2.1   Knowing the Orientation

The first case we study will be the simpler one among the examples. We imagine (1) from Table 7.2 to be the p specification which says that the agent believes v to be uniformly distributed on the interval [0, 10], $h = 6$ and $\theta = 0$. This is a complete specification, in the sense that a unique joint distribution is provided. Moreover, owing to the exact knowledge that the robot has about its orientation, it is very certain on when the sonar would reach WALLCLOSE and when it would stop at WALLFAR, *viz.* the situations where $v \in [1, 2]$ or $v \in [3, 4]$ or $v \in [7, 8]$ are the only epistemically possible ones where *Blocked* will not hold. Therefore, the agent initially believes v to be uniform, as shown in Fig. 7.2, but after sensing 5.9, v values in the gaps will be considered with high probability (and equally likely) while the remaining v values will be given low p values.

Example 7.1 Let $\mathcal{D}$ be a basic action theory that includes the sentences in Tables 7.1 and 7.2. Here are some properties of the basic action theory stated more formally:

1. $\mathcal{D} \models Bel(v \in [3, 4.57], S_0) = 0.157$
   To see how this number is obtained, let us first expand *Bel* to obtain:

Table 7.2 Certainty about θ

1. $p(\iota, S_0) =$	$\begin{cases} 0.1 & \text{if}(h = 6 \wedge v \in [0, 10] \wedge \theta = 0)[\iota] \\ 0 & \text{otherwise} \end{cases}$

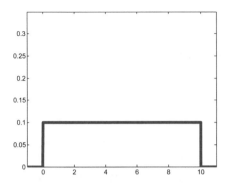
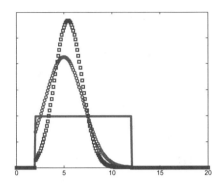

Fig. 7.2 Beliefs initially and after sensing 5.9 when $\theta = 0$

$$\frac{1}{\gamma} \int_x \int_y \int_z \begin{cases} p(\iota, S_0) & \exists \iota (h = x \wedge v = y \wedge \theta = z \wedge v \in [3, 4.57])[\iota] \\ 0 & \text{otherwise} \end{cases}$$

By means of the p-specification in $\mathcal{D}_0$, this simplifies to:

$$\frac{1}{\gamma} \int_x \int_y \int_z \begin{cases} 0.1 & \exists \iota (h = x \wedge v = y \wedge \theta = z \wedge \ldots)[\iota] \\ 0 & \text{otherwise} \end{cases}$$

where the ellipsis stands for:

$$(h = 6 \wedge v \in [0, 10] \wedge \theta = 0) \wedge (v \in [3, 4.57])$$

Intuitively, for the numerator of *Bel*, we are to integrate a function $q(x, y, z)$ (where x corresponds to the fluent h, y corresponds to the fluent v and z corresponds to the fluent θ) that is 0.1 when $y \in [3, 4.57]$ and 0 otherwise. This equals $0.157/\gamma$. (The simplified mathematical expressions in this example and the ones below can be calculated using any software with numerical integration capabilities. We often write $t \approx u$ when the calculation for an expression t gives u after truncating the resulting value to two significant digits.) The normalization factor γ, analogously, is shown to be:

$$\int_x \int_y \int_z \begin{cases} 0.1 & \exists \iota \, (h = x \wedge v = y \wedge \theta = z \wedge h = 6 \wedge v \in [0, 10] \wedge \theta = 0)[\iota] \\ 0 & \text{otherwise} \end{cases}$$

Here, $v \in [3, 4.57]$ from the numerator is dropped (i.e., replaced by *true*). It is easy to see that $\gamma = 1$ since y is integrated for all values.

2. *Bel*$(v \in [3, 4], do(sonar(5.9), S_0)) \approx 0.33$

   We do the expansion of *Bel* in detail for this one too. We have:

$$\frac{1}{\gamma} \int_x \int_y \int_z \begin{cases} p(do(sonar(5.9), \iota), do(sonar(5.9), S_0)) & \text{if } \exists \iota(\ldots)[\iota] \\ 0 & \text{otherwise} \end{cases}$$

where the ellipsis stands for

$$h = x \wedge v = y \wedge \theta = z \wedge v(do(sonar(5.9), now)) \in [3, 4].$$

By means of the specification in $\mathcal{D}_0$ and the l-axiom for the distance sensor, we obtain:

$$\frac{1}{\gamma} \int_x \int_y \int_z \begin{cases} 0.1 \cdot \mathcal{N}(\delta + \lambda - 5.9; 0, 1) & \text{if } \exists \iota(\ldots \wedge \psi)[\iota] \\ 0.1 \cdot \mathcal{N}(\delta - 5.9; 0, 1) & \text{if } \exists \iota(\ldots \wedge \neg\psi)[\iota] \\ 0 & \text{otherwise} \end{cases}$$

where, the ellipsis stands for

$$h = x \wedge v = y \wedge \theta = z \wedge$$
$$h = 6 \wedge v \in [0, 10] \wedge \theta = 0 \wedge$$
$$v(do(sonar(5.9), now)) \in [3, 4];$$

and ψ denotes

$$(v + \tan \theta \in [1, 2]) \vee (v + \tan \theta \in [3, 4]) \vee (v + \tan \theta \in [7, 8]).$$

The idea is simple. First, from (1), those initial situations where $h = 6$, $\theta = 0$ and $v \in [0, 10]$ are the only ones with non-zero p values. By means of (P3), for various real values of y, which is the variable corresponding to v, we will be ranging over all the initial situations with non-zero p values. Next, since we are interested in the belief in $v \in [3, 4]$, as in the previous item, we give all other successor situations a density of 0 when calculating the numerator. Third, we note that when *Blocked* holds (tested using ψ), from (7) and (P2), p values get multiplied by $\mathcal{N}(\delta - 5.9; 0, 1)$, and when not, p values get multiplied by $\mathcal{N}(\delta + \lambda - 5.9; 0, 1)$. Analogously, for γ, we derive a similar formula by replacing $v(do(sonar(5.9), now)) \in [3, 4]$ in the conditional expressions by *true*. As a final simplification in the numerator, because $\tan \theta = 0$, the second condition in the case statement (with $\neg\psi$) is not satisfiable, and so we get:

$$\frac{1}{\gamma} \int_3^4 0.1 \cdot \mathcal{N}(0.1; 0, 1) \approx 0.33.$$

7.2.2 Uncertainty About the Orientation

We now consider a more interesting p specification determined by the orientation. The p we are thinking of is the one specified in Table 7.3. Here, $h = 6$, v is uniformly distributed as before, and independently θ is normally distributed around 0 with a variance of 9. This too is a complete specification, in the sense that there is a unique joint distribution corresponding to the p axiom.

Consider for the moment what would happen after sensing once. Unlike in Fig. 7.2, there is uncertainty regarding θ, which means that sensing (say) 5.9 will not imply full confidence in v being in $[1, 2] \cup [3, 4] \cup [7, 8]$. Indeed, as discussed earlier, even for v values less than 1, the orientation may cause the sonar to sense WALLCLOSE. Moreover, a larger range of θ values may cause the sonar to sense WALLCLOSE in the $[3, 4]$ interval rather than the $[1, 2]$ interval due to its lack of wall obstructions, causing a belief density change as shown in Fig. 7.3. After moving (say) 2 units and sensing values closer to $\lambda + \delta$ will lead to a more definite localization, as also shown in Fig. 7.3.

Example 7.2 Let $\mathcal{D}$ be a basic action theory that includes the sentences in Tables 7.1 and 7.3. Here are some properties of this second basic action theory:

1. $\mathcal{D} \models Bel(v \in [3, 4.57], S_0) = 0.157$
   We are integrating under the same conditions initially as in the previous example, except that we now have:

$$\frac{1}{\gamma} \int_x \int_y \int_z \begin{cases} 0.1 \cdot \mathcal{N}(z; 0, 1) & \text{if } \exists \iota. \ (h = x \wedge v = y \wedge \theta = z \wedge \ldots)[\iota] \\ 0 & \text{otherwise} \end{cases}$$

   where the ellipsis stands for

$$h = 6 \wedge v \in [0, 10] \wedge v \in [3, 4.57]$$

   and where the distribution for θ is also considered in the density expression for initial situations. Here, because we are integrating θ for all values, we obtain $\gamma = 1$ as usual, and so the above expression leads to 0.157.

2. $\mathcal{D} \models Bel(v \in [3, 4], do(sonar(5.9), S_0)) \approx 0.31$
   $\mathcal{D} \models Bel(v \in [2.8, 4.2], do(sonar(5.9), S_0)) \approx 0.33$

Table 7.3 Uncertainty about θ

1. $\quad p(\iota, S_0) =$	$\begin{cases} 0.1 \times \mathcal{N}(\theta(\iota); 0, 9) & \text{if}(h = 6 \wedge v \in [0, 10])[\iota] \\ 0 & \text{otherwise} \end{cases}$

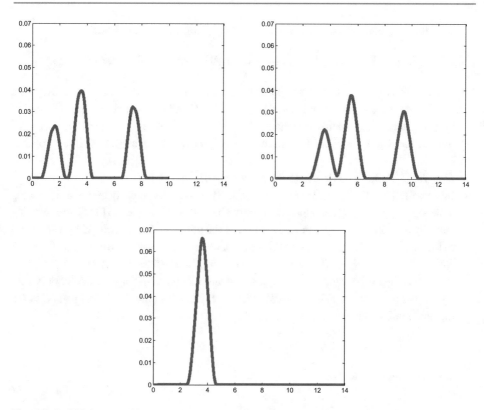

Fig. 7.3 Belief change with normally distributed θ: after sensing 5.9, moving 2 units after that, and sensing 5.83 finally

It is worth developing this in detail and contrasting it with what we had in the previous example. Picking the first entailment, *Bel* can be seen to expand as:

$$\frac{1}{\gamma} \int_x \int_y \int_z 0.1 \cdot \mathcal{N}(z; 0, 9) \cdot \begin{cases} \mathcal{N}(0.1; 0, 1) & \text{if } \exists \iota(\ldots \wedge \psi)[\iota] \\ \mathcal{N}(4.9; 0, 1) & \text{if } \exists \iota(\ldots \wedge \neg \psi)[\iota] \\ 0 & \text{otherwise} \end{cases}$$

where (analogously) the ellipsis stands for:

$$h = x \wedge \theta = z \wedge v = y \wedge$$
$$h = 6 \wedge v \in [0, 10] \wedge$$
$$v(do(sonar(5.9), now)) \in [3, 4]$$

and ψ is (exactly as before):

$$(v + \tan \theta \in [1, 2]) \vee (v + \tan \theta \in [3, 4]) \vee (v + \tan \theta \in [7, 8]).$$

Note the simplification of the *l*-values for the sensing action as follows:

$$\mathcal{N}(\delta + \lambda - 5.9; 0, 1) = \mathcal{N}(0.1; 0, 1), \mathcal{N}(\delta - 5.9; 0, 1) = \mathcal{N}(4.9; 0, 1).$$

What is interesting about *Bel*'s expansion here is that since $\theta \neq 0$, the sensor may read $\delta + \lambda$ even if the robot is not located in $[1, 2]$, $[3, 4]$ and $[7, 8]$. This accounts for belief in (say) exactly $[3, 4]$ being less than 1/3, which is different from the previous example. Indeed, the degree of belief in a slightly larger interval, such as $[2.8, 4.2]$, approaches 1/3.

3. $\mathcal{D} \models Bel(v \in [3, 4], do([sonar(5.9), move(2, 90)], S_0)) \approx 0.31$
   Intuitively, the belief in $[1, 2]$ after sensing 5.9, which is also slightly less than 1/3 owing to the open door at $[1, 2]$ and the uncertainty about θ, is transferred to $[3, 4]$ after moving laterally by 2 units. (That is, the new values of the points in the interval $[1, 2]$ correspond to the interval $[3, 4]$ and their densities do not change over a noise-free action.)

4. $\mathcal{D} \models Bel(v \in [3, 4], do([sonar(5.9), move(2, 90), sonar(5.83)], S_0)) \approx 0.96$
   After sensing a value close to $\lambda + \delta$, moving and sensing $\lambda + \delta$ again, the robot is very confident about the $[3, 4]$ interval. We will not expand *Bel* completely but just point out that the density function is

$$0.1 \cdot \mathcal{N}(z; 0, 9) \cdot \mathcal{N}(\delta + \lambda - 5.9; 0, 1) \cdot N(\delta + \lambda - 5.83; 0, 1)$$

at initial situations where:

$$h = 6 \wedge v \in [0, 10] \wedge$$
$$(v + \tan \theta \in [1, 2] \vee v + \tan \theta \in [3, 4] \vee v + \tan \theta \in [7, 8]) \wedge$$
$$(v + \tan \theta + 2 \in [1, 2] \vee v + \tan \theta + 2 \in [3, 4] \vee$$
$$v + \tan \theta + 2 \in [7, 8]).$$

Roughly speaking, these situations are those that support the observations of 5.9 and 5.83 in the best possible way. Note that, for the second sensing action, we need to test whether the incremented value of v after $move(2, 90)$ is within a gap. It is not hard to see that when v is in the vicinity of $[3, 4]$, we would easily satisfy these constraints, which then has the intended effect.

7.3   Notes

As seen in much of the work in cognitive robotics [138, 190], a logical language like the situation calculus allows for non-trivial action specifications, including, for example, context-dependent prerequisites and effects. In earlier chapters, we have demonstrated how such actions can affect probability distributions in interesting ways, such as transforming continuous distributions to mixed ones, and how the language leads itself for reasoning about past and future events, among others.

Most significantly, in comparison to standard (non-logical) probabilistic formalisms, the advantages of the situation calculus are perhaps most evident in terms of what is allowed in the initial specification of the p fluent. The two examples used in this chapter were comparable to unique joint probability distributions, which are standard.[1] But that is not the case for one of the form:

$$\forall \iota (p(\iota, S_0) = \mathcal{U}(v; 0, 10)[\iota]) \vee \forall \iota (p(\iota, S_0) = \mathcal{U}(v; 3, 13)[\iota])$$

This says that the agent believes v to be uniformly distributed on $[0, 10]$ or on $[3, 13]$, without being able to say which. (That is, the $\mathcal{U}$ term is an abbreviation for the mathematical formula defining a uniform density.) As one would expect (in logic), appropriate beliefs will still be entailed. For example:

- initially, it will follow that the robot is certain that $v \notin [30, 40]$, and will believe that $v \in [3, 10]$ with a probability of 0.7;
- if the robot has sensors to indicate that it is well within (say) the range of $[7, 8]$, after a few sensor readings, the disjunctive uncertainty about v will no longer be significant.

Much weaker specifications are possible still, where the modeler may leave the nature of the distribution of some fluents completely open, which would correspond more closely to incomplete information in the usual categorical sense [57], among others. As an example, the sentence above in a language with even one other fluent, say h, would mean that the agent has no information about the initial distribution of h. All of these are admitted in the framework.

We note that localization has been addressed using a number of algorithmic techniques for more than two decades in the robotics literature [51, 63, 219]. Much of the results study algorithmic tradeoffs, such as investigating sampling-based techniques, approximating domains with Gaussian distributions, and so on. So, with regards to the underlying formal characterization, almost all of these are based on Bayesian conditioning.

Our objective was not be to compete with these techniques; in fact, the chapter did not concern itself with algorithms at all. Rather, we wanted to show how localization can be understood as part of a larger effort in a single logical framework [190]. In the same vein, this chapter is not to be seen as suggesting that one should use a theorem prover to compute such beliefs, or even that computing localization must be performed with such an axiomatization. It is only showing that the language is representational adequate, offering a formal perspective on the problem. In practice, we may need to consider hybrid options that amalgamate high-level logical reasoning and low-level, sensory data computations; see, for

---

[1] The example from this chapter is adapted from [219]. More general formalizations can be considered; see, for example, [84, 143] for logical representations of richer geometric theories.

example, the myriad of proposals in [34, 38, 75, 119, 120, 175, 176]. We will have more to say about this in the final chapter.

With matters of representational adequacy addressed, let us now turn to the question of how to address projection effectively.

Regression and Progression

8

> *Life can only be understood backwards; but it must be lived forwards.*
> —*Soren Kierkegaard*

While the situation calculus is a powerful representation language, the task of how to effectively reason about actions is still to be addressed in the context of probabilistic beliefs. In essence, this would correspond to a more involved projection problem, where the state of knowledge, actions and sensing are mixtures of logical and probabilistic assertions. More precisely, while changing degrees of belief do indeed emerge as *logical entailments* of the given action theory, no procedure is given for computing these entailments. On closer examination, in fact, this is a two-part question:

(i) How do we effectively reason about beliefs in a particular state?
(ii) How do we effectively reason about belief state evolution and belief change?

In the simplest case, part (i) puts aside acting and sensing, and considers reasoning about the *initial* state only, which is then the classical problem of (first-order) probabilistic inference. We do not attempt to do a full survey here, but as discussed in previous chapters, this has received a lot of attention.

This chapter is about part (ii). Addressing this concern would not only aid planning algorithms, but also has a critical bearing on the assumptions made about the domain for tractability purposes. For example, if the initial state supports a decomposed representation of the distribution, can we expect the same after actions? In the exception of very limited cases such as Kalman filtering that harness the conjugate property of Gaussian processes, the situation is discouraging. In fact, even in the slightly more general case of Dynamic Bayesian Networks, which are in essence atomic propositions, if one were to assume that

© The Author(s), under exclusive license to Springer Nature Switzerland AG 2023   117
V. Belle, *Toward Robots That Reason: Logic, Probability & Causal Laws*,
Synthesis Lectures on Artificial Intelligence and Machine Learning,
https://doi.org/10.1007/978-3-031-21003-7_8

state variables are independent at time 0, they can become fully correlated after a few steps. Dealing with complex actions, incomplete specifications and mixed representations, therefore, is significantly more involved.

As discussed previously, in the reasoning about actions literature, where the focus is on qualitative (non-probabilistic) knowledge, there are two main solutions to projection (Sect. 3.2.5): regression and progression. Both of these have proven enormously useful for the design of logical agents, essentially paving the way for cognitive robotics. Roughly, regression reduces a query about the future to a query about the initial state. Progression, on the other hand, changes the initial state according to the effects of each action and then checks whether the formula holds in the updated state. In this chapter, we show how both of these generalize in the presence of degrees of belief, noisy acting and sensing. Our results allow for both discrete and continuous probability distributions to be used in the specification of beliefs and dynamics, that leverage the extension of the situation calculus to mixed discrete-continuous domains, discussed in a previous chapter.

To elaborate on the regression result, we show that it is general, not requiring (but allowing) structural constraints about the domain, nor imposing (but allowing) limitations to the family of actions. Regression derives a mathematical formula, using term and formula *substitution* only, that relates belief after a sequence of actions and observations, even when they are noisy, to beliefs about the initial state. That is, among other things, if the initial state supports efficient factorizations, regression will maintain this advantage no matter how actions affect the dependencies between state variables over time. Going further, the formalism will work seamlessly with discrete probability distributions, probability densities, and perhaps most significantly, with difficult combinations of the two.

To see a simple example of what goal regression does, imagine a robot facing a wall and at a certain distance h to it, as in Fig. 3.2. The robot might initially believe h to be drawn from a uniform distribution on $[2, 12]$. Assume the robot moves away by 2 units and is now interested in the belief about $h \leq 5$. Regression would tell the robot that this is equivalent to its initial beliefs about $h \leq 3$ which here would lead to a value of 0.1. To see a nontrivial example, imagine now the robot is also equipped with a sonar unit aimed at the wall, that adds Gaussian noise with mean μ and variance σ^2. After moving away by 2 units, if the sonar were now to provide a reading of 8, then regression would derive that belief about $h \leq 5$ is equivalent to

$$\frac{1}{\gamma} \int_2^3 0.1 \times \mathcal{N}(6 - x; \mu, \sigma^2) \, dx.$$

where γ is the normalization factor. Essentially, the posterior belief about $h \leq 5$ is reformulated as the product of the prior belief about $h \leq 3$ and the likelihood of $h \leq 3$ given an observation of 6. (That is, observing 8 after moving away by 2 units is related here to observing 6 initially.)

We believe the broader implications of this result are two-fold. On the one hand, as we show later, simple cases of belief state evolution, as applicable to conjugate distributions for example, are special cases of regression's backward chaining procedure. Thus, regression could serve as a *formal basis* to study probabilistic belief change w.r.t. limited forms of actions. On the other hand, our contribution can be viewed as a methodology for combining actions with recent advances in probabilistic inference, because reasoning about actions reduces to reasoning about the initial state.

To elaborate on the progression result, it has been argued that for long-lived agents like robots, continually updating the current view of the state of the world, is perhaps better suited. Lin and Reiter show that progression is always second-order definable, and in general, it appears that second-order logic is unavoidable. However, Lin and Reiter also identify some first-order definable cases by syntactically restricting situation calculus basic action theories, and since then, a number of other special cases have been studied.

While Lin and Reiter intended their work to be used on robots, one criticism leveled at their work, and indeed at much of the work in cognitive robotics, is that the theory is far removed from the kind of continuous uncertainty and noise seen in typical robotic applications. What exactly filtering mechanisms (such as Kalman filters) have to do with Lin and Reiter's progression has gone unanswered, although it has long been suspected that the two are related.

Our result remedies this situation. However, as we discuss later, progression in stochastic domains is complicated by the fact that actions can transform a continuous distribution to a mixed one. To obtain a closed-form result, we introduce a property of basic action theories called *invertibility*, closely related to invertible functions in real analysis. We identify syntactic restrictions on basic action theories that guarantee invertibility. For our central result, we show a first-order *progression* of degrees of belief against noise in effectors and sensors for action theories that are invertible.

8.1   Regression for Discrete Domains

For both regression and progression, it will be convenient to treat successor state and like-lihood axioms equationally, as we have for p-axioms in our examples. In particular, henceforth, successor state axioms are taken to be in the form:

$$f(do(a, s)) = E_f(a)[s]. \tag{8.1}$$

Likewise, the action theory $\mathcal{D}$ is assumed to contain for each sensor $sense(\vec{x})$ that measures a fluent f, an axiom of the form:

$$l(sense(\vec{x}), s) = Err_{sense}(\vec{x}, f(s)),$$

where $Err_{sense}(u_1, u_2)$ is some expression with only two free variables u_1 and u_2, both numeric.[1]

---

[1] This captures the idea that the error model of a sensor measuring f depends only on the true value of f, and is independent of other factors. In a sense this follows the Bayesian model that conditioning

As before, noise-free physical actions are given a likelihood of 1. Noisy physical actions will be treated in a subsequent section.

We now investigate a computational mechanism for reasoning about beliefs after actions. In this section, we focus on discrete domains, where a *weight*-based notion of belief is appropriate. Domains with both discrete and continuous variables are reserved for the next section.

Recall that, with projection, given a basic action theory $\mathcal{D}$, a sequence of actions δ, we might want to determine whether a formula ϕ *holds* after executing δ starting from S_0:

$$\mathcal{D} \models \phi[do(\delta, S_0)]. \tag{8.2}$$

When it comes to beliefs, and in particular how that changes after acting and sensing, we might be interested in *calculating* the degrees of belief in ϕ after δ: find a real number r such that

$$\mathcal{D} \models Bel(\phi, do(\delta, S_0)) = r. \tag{8.3}$$

The obvious method for answering this is to translate both the action theory and the query into a predicate logic formula, and apply regression as introduced for non-probabilistic variants. This approach, however, presents serious problems, both from a computational as well as a readability viewpoint. Recall that belief formulas expand into a complicated formula mentioning initial and successor situations. Likewise, sums (and integrals in the continuous case) expand to involved second-order formulas.

Therefore, we now introduce a *regression* operator to simplify both (8.2) and (8.3) to queries about $Bel(\phi, S_0)$, over arithmetic expressions, for which standard probabilistic reasoning methods can be applied. For this purpose, in the sequel, Bel is treated as a *special syntactic operator* rather than as an abbreviation for other formulas. To see a simple example of the procedure, imagine the robot is interested in the probability of $h = 7$, given (6.2), after reading 5 from a sonar:

$$Bel(h = 7, do(sonar(5), S_0)) \tag{8.4}$$

Suppose we take the sonar's model to be the following:

$$l(sonar(z), s) = \text{IF } |h(s) - z| \leq 1 \text{ THEN } 1/3 \text{ ELSE } 0 \tag{8.5}$$

Then (8.4) should be 0 by Bayesian conditioning because the likelihood of the true value being 7 given an observation of 5 is 0. Regression would reduce the term (8.4) to one over initial priors:

---

on a random variable f is the same as conditioning on the event of observing f. But this is not required in general in the situation calculus, as illustrated in previous chapters, an issue we ignore for this chapter. As before, we assume that physical actions have trivial sensing values, and that sensing actions do not affect physical properties. Physical actions with non-trivial sensing axioms are to be treated as two separate actions, the first capturing the physical effects and the second capturing the sensing ones.

$$\frac{1}{\gamma} \sum_{x \in \{2,\ldots,11\}} Err(5, x) \times Bel(h = x \wedge h = 7, S_0) \tag{8.6}$$

where Err is the error model from (8.5). By the condition inside Bel, the only valid value for x is 7 for which the prior is 0.1 but $Err(5, 7)$ is 0. Thus, $(8.4) = (8.6) = 0$. In general, regression is a *recursive* procedure that works iteratively over a sequence of actions discarding one action at a time, and it can be utilized to measure any logical property about the variables, e.g. $2\pi \cdot h < 12, h/fuel \leq mileage$, etc.

Formally, regression operates at two levels, which is different to the treatment for *Knows*. At the formula level, we introduce an operator $\mathcal{R}$ for regressing formulas, which over equality literals sends the individual terms to an operator $\mathcal{T}$ for regressing terms. These operators proceed by mutual recursion. The fundamental objective of these operators is eliminate *do* symbols. The end result, then, is to transform any expression whose situation term is a successor of S_0, say $do([a_1, a_2], S_0)$, to one about S_0 only, at which point $\mathcal{D}_0$ is all that is needed. As hinted earlier, these operators treat $Bel(\phi, s)$ as though they are special sorts of terms. Throughout the presentation, we assume that the inputs to these operators do not quantify over all situations.

Definition 8.1 For any term t, we inductively define $\mathcal{T}[t]$:

1. If t is situation-independent (e.g. $x, \pi^{2/3}$) then $\mathcal{T}[t] = t$.
2. $\mathcal{T}[g(t_1, \ldots, t_k)] = g(\mathcal{T}[t_1], \ldots, \mathcal{T}[t_k])$,
   where g is any non-fluent function (e.g. $\times, +, \mathcal{N}$).
3. For a fluent function f, $\mathcal{T}[f(s)]$ is defined inductively

   (a) if s is of the form $do(a, s')$ then
       $\mathcal{T}[f(s)] = \mathcal{T}[E_f(a)[s']]$
   (b) else $\mathcal{T}[f(s)] = f(s)$

   where, in (a), we use the instance of the RHS of the successor state axiom w.r.t. a, assumed to be in the form above.
4. $\mathcal{T}[Bel(\phi, s)]$ is defined inductively:

   (a) if s is of the form $do(a, s')$ and a is a noise-free physical action, then
       $\mathcal{T}[Bel(\phi, s)] = \mathcal{T}[Bel(\psi, s')]$
       where ψ is $Poss(a, now) \supset \mathcal{R}[\phi[do(a, now)]]$.
   (b) if s is of the form $do(a, s')$ and a is a sensing action $sense(z)$ such that $l(sense(z), s) = Err(z, f_i(s))$ is in $\mathcal{D}$ then
       $\mathcal{T}[Bel(\phi, s)] =$
       $\frac{1}{\gamma} \sum_{x_i} Err(z, x_i) \times \mathcal{T}[Bel(\psi, s')]$

where ψ is $Poss(a, now) \supset \phi \wedge f_i(now) = x_i$, and γ is the normalization factor and is the same expression as the numerator but ϕ replaced by *true*.

(c) else $T[Bel(\phi, s)] = Bel(\phi, s)$.

Definition 8.2 For any formula ϕ, we define $\mathcal{R}[\phi]$ inductively:

1. $\mathcal{R}[t_1 = t_2] = (T[t_1] = T[t_2])$
2. $\mathcal{R}[G(t_1, \ldots, t_k)] = G(T[t_1], \ldots, T[t_k])$
   where G is any non-fluent predicate (e.g. =, <).
3. When ψ is a formula, $\mathcal{R}[\neg\psi] = \neg\mathcal{R}[\psi]$,
   $\mathcal{R}[\forall x \psi] = \forall x \mathcal{R}[\psi], \mathcal{R}[\exists x \psi] = \exists x \mathcal{R}[\psi]$.
4. When ψ_1 and ψ_2 are formulas,
   $\mathcal{R}[\psi_1 \wedge \psi_2] = \mathcal{R}[\psi_1] \wedge \mathcal{R}[\psi_2]$,
   $\mathcal{R}[\psi_1 \vee \psi_2] = \mathcal{R}[\psi_1] \vee \mathcal{R}[\psi_2]$.
5. $\mathcal{R}[Poss(A(\vec{t}), s)] = \mathcal{R}[\Pi_A(\vec{t}, s)]$,
   where the instance of the RHS of the precondition axiom w.r.t. $A(\vec{t})$ replaces the atom.

This completes the definition of T and $\mathcal{R}$. We now go over the justifications for the items, starting with the operator T. In item 1, non-fluents simply do not change after actions. In item 2, T operates over sums and products in a modular manner. In item 3, provided there are remaining *do* symbols, the physics of the domain determines what the conditions must have been in the previous situation for the current value to hold. In item 4, if there is a remainder physical action, part (a) says that belief in ϕ after actions is simply the prior belief about the regression of ϕ, contingent on action executability. Part (b) says that the belief about ϕ after observing z for the true value of f_i is the prior belief for all possible values x_i for f_i that agree with ϕ, times the likelihood of f_i being x_i given z. The appropriateness of parts (a) and (b) depend on the fact that physical actions do not have any sensing aspect, while sensing actions do not change the world. Part (c) simply says that T stops when no *do* symbols appear in s. We proceed now with the justifications for $\mathcal{R}$. Over equality atoms, $\mathcal{R}$ separates the terms of the equality and sends them to T. Likewise, over non-fluent predicates. Also, $\mathcal{R}$ simplifies over connectives in a straightforward way. When *Poss* is encountered, preconditions take its place.

The main result for $\mathcal{R}$ regarding projection is:

Theorem 8.3 *Suppose* $\mathcal{D}$ *is any action theory,* ϕ *any situation-suppressed formula possibly mentioning Bel, and* δ *any action sequence:*

$$\mathcal{D} \models \phi[do(\delta, S_0)] \quad \textit{iff} \quad \mathcal{D}_0 \cup \mathcal{D}_{una} \models \mathcal{R}[\phi[do(\delta, S_0)]]$$

where $\mathcal{D}_{una}$ *is the unique name assumption and* $\mathcal{R}[\phi[do(\delta, S_0)]]$ *mentions only a single situation term,* S_0.

Here, $\mathcal{D}_{una}$ is only needed to simplify action terms, e.g. from $move(4) = move(z)$, $\mathcal{D}_{una}$ infers $z = 4$.

The readers may notice many parallels between this regression operator, and the one for *Knows*, which should not be surprising as the new operator is a generalization of the previous one.

Now when our goal is to explicitly compute the degrees of belief in the sense of (8.3), we have the following property for $\mathcal{T}$, which follows as a corollary from the above theorem:

Theorem 8.4 *Let $\mathcal{D}$ be as above, ϕ any situation-suppressed formula and δ any sequence of actions. Then:*

$$\mathcal{D} \models Bel(\phi, do(\delta, S_0)) = \mathcal{T}[Bel(\phi, do(\delta, S_0))]$$

where $\mathcal{T}[Bel(\phi, do(\delta, S_0))]$ is a term about S_0 only.

Theorem 8.4 essentially shows how belief about trajectories is computable using beliefs about S_0 only. Note that, since the result of $\mathcal{T}$ is a term about S_0, no sentence outside of $\mathcal{D} - \mathcal{D}_0$ is needed. We now illustrate regression with examples. Using Theorem 8.4, we reduce beliefs after actions to initial ones. At the final step, standard probabilistic reasoning is applied to obtain the end values.

Example 8.5 Let $\mathcal{D}$ contain the union of (6.1), (6.2), and (8.5) as the likelihood axiom for the sensor.

Then the following equality expressions are entailed by $\mathcal{D}$:

1. $Bel(h = 10 \vee h = 11, S_0) = 0.2$
   $Bel(h \leq 9, S_0) = 0.8$
   Terms about S_0 are unaffected by $\mathcal{T}$. So this amounts to inferring probabilities using $\mathcal{D}_0$.
2. $Bel(h = 11, do(move(1), S_0))$
   $= \mathcal{T}[Bel(h = 11, do(move(1), S_0))]$
   $= \mathcal{T}[Bel(\underline{\mathcal{R}[(h = 11)[do(move(1), now)]]}, S_0)]$ (i)
   $= \mathcal{T}[Bel(\underline{\mathcal{T}[h(do(move(1), now))] = \mathcal{T}[11]}, S_0)]$ (ii)
   $= \underline{\mathcal{T}[Bel(max(0, h - 1) = 11, S_0)]}$ (iii)
   $= Bel(max(0, h - 1) = 11, S_0)$ (iv)
   $= 0$

First, since action preconditions are all true, *Poss* is ignored everywhere. We underline to emphasize the expressions undergoing transformations. We begin always by applying $\mathcal{T}$ to the main term, in this case getting (i), by means of $\mathcal{T}$'s item 4(a). Next, $\mathcal{R}$'s item 1 is applied in (ii). While $\mathcal{T}[11] = 11$ by $\mathcal{T}$'s item 1, for $\mathcal{T}[h(do(move(1), now))]$ we use item 3 and (6.1) to get:

$$\mathcal{T}[max(0, h(now) - 1)] = max(0, h(now) - 1)$$

which is substituted in (ii) to give (iii). Finally, $\mathcal{T}$'s item 4(c) yields (iv), which is a belief term about S_0. Now the only valid value for h in (iv) is 12, but for $h = 12$ the robot has a belief of 0 initially.

3. $Bel(h \leq 5, do(sonar(5), S_0))$

$$= \frac{1}{\gamma} \sum_{x \in \{2,...,11\}} Err(5, x) \times \underline{\mathcal{T}[Bel(h = x \wedge h \leq 5, S_0)]} \tag{i}$$

$$= \frac{1}{\gamma} \sum_{x \in \{2,...,11\}} Err(5, x) \times Bel(h = x \wedge h \leq 5, S_0) \tag{ii}$$

$$= \frac{1}{\gamma} \left(\frac{1}{3} \cdot Bel(h = 4 \wedge h \leq 5, S_0) \right.$$
$$+ \frac{1}{3} \cdot Bel(h = 5 \wedge h \leq 5, S_0) \tag{iii}$$
$$\left. + \frac{1}{3} \cdot Bel(h = 6 \wedge h \leq 5, S_0) \right)$$

$$= \frac{1}{\gamma} \left(\frac{1}{3} \cdot Bel(h = 4, S_0) + \frac{1}{3} \cdot Bel(h = 5, S_0) \right) \tag{iv}$$

$$= \frac{1}{\gamma} \cdot \frac{2}{30}$$

$$= 2/3$$

where $Err(5, x)$ is the model from (8.5). First, $\mathcal{T}$'s item 4(b) yields (i), and then item 4(c) yields (ii). Since $Err(5, x)$ is non-zero only for $x \in \{4, 5, 6\}$, (ii) is simplified to (iii) and (iv) resulting in $1/15 \cdot 1/\gamma$. We calculate γ as follows:

$$= \sum_{x \in \{2,...,11\}} Err(5, x) \times \mathcal{T}[Bel(h = x \wedge true, S_0)] \tag{i$'$}$$

$$= \sum_{x \in \{2,...,11\}} Err(5, x) \times Bel(h = x, S_0) \tag{ii$'$}$$

$$= 3/30.$$

8.2 Regression for General Domains

We now generalize regression for domains with discrete and continuous variables, for which a *density*-based notion of belief is appropriate. (Physical actions are still noise-free for this section.) The main issue is that when formulating posterior beliefs after sensing, something like Definition 8.1's item 4(b) will not work. This is because over continuous spaces the belief about any individual point is 0. Therefore, we will be unpacking belief in terms of the density function, i.e. in terms of P. These $P(\vec{x}, \phi, s)$ terms, which will now also be treated as special sorts of syntactic terms, are separately regressed. (Of course, the regression of weight-based belief can be approached on similar lines.) Recall that $P(\vec{x}, \phi, S_0)$ is simply the *density* of an initial world (where $f_i = x_i$) satisfying ϕ. Formally, term regression $\mathcal{T}$ is defined as follows:

Definition 8.6 For any term t, we inductively define:

1, 2 and 3 as before (Definition 8.1).

4. $\mathcal{T}[P(\vec{x}, \phi, s)]$ is defined inductively:

   (a) if s is of the form $do(a, s')$ and a is a physical action then
   $$\mathcal{T}[P(\vec{x}, \phi, s)] = \mathcal{T}[P(\vec{x}, \psi, s')]$$
   where ψ is $Poss(a, now) \supset \mathcal{R}[\phi[do(a, now)]]$.

   (b) if s is of the form $do(a, s')$ and a is a sensing action $sense(z)$ such that $l(sense(z), s) = Err(z, f_i(s))$ is in $\mathcal{D}$, then:
   $$\mathcal{T}[P(\vec{x}, \phi, s)] = Err(z, x_i) \times \mathcal{T}[P(\vec{x}, \psi, s')]$$
   where ψ is $Poss(a, now) \supset \phi \wedge f_i(now) = x_i$.

   (c) else $\mathcal{T}[P(\vec{x}, \phi, s)] = P(\vec{x}, \phi, s)$.

5. $\mathcal{T}[Bel(\phi, s)] = \dfrac{1}{\gamma} \displaystyle\int_{\vec{z}} \mathcal{T}[P(\vec{z}, \phi, s)]$.

   $\mathcal{R}$ for formulas is defined as before.

It is worth observing that there is no summation (or integration) symbol when applying $\mathcal{T}$ over noisy sensors because $\mathcal{T}$ over *Bel* expands it first as the integral over the unnormalized density expression $P(\vec{z}, \phi, s)$. In contrast, previously $\mathcal{T}$'s application over a *Bel* term in Definition 8.1 did not modify the term.

With this new definition, the desired property still holds:

Theorem 8.7 *Let $\mathcal{D}$ be any action theory, ϕ any situation-suppressed formula and δ any action sequence. Then*

$$\mathcal{D} \models Bel(\phi, do(\delta, S_0)) = \mathcal{T}[Bel(\phi, do(\delta, S_0))]$$

where $\mathcal{T}[Bel(\phi, do(\delta, S_0))]$ is a term about S_0 only.

Example 8.8 Consider the following *continuous* variant of the robot example. Imagine a continuous uniform distribution for the true value of h, as provided by (6.2). Suppose the sonar has the following error profile:

$$l(sonar(z), s) = \text{IF } z \geq 0$$
$$\text{THEN } \mathcal{N}(z - h(s); 0, 4) \tag{8.7}$$
$$\text{ELSE } 0$$

which says the difference between a nonnegative reading and the true value is normally distributed with mean 0 and variance 4. (A mean of 0 implies there is no systematic bias.) Now, let $\mathcal{D}$ be any action theory that includes (6.1), (6.2) and (8.7). Then the following equalities are entailed by $\mathcal{D}$:

1. $Bel(h = 3 \vee h = 4, S_0) = 0$,
   $Bel(4 \leq h \leq 6, S_0) = 0.2$
   $\mathcal{T}$ does not change terms about S_0. Here, for example, the second belief term equals $\int_4^6 0.1 dx = 0.2$.

2. $Bel(h \geq 11, do(move(1), S_0))$

$$= \frac{1}{\gamma} \int_{x \in \mathbb{R}} \mathcal{T}[\, P(x, \underline{h \geq 11}, do(move(1), S_0))\,] \tag{i}$$

$$= \frac{1}{\gamma} \int_{x \in \mathbb{R}} \mathcal{T}[P(x, \underline{\mathcal{R}[\psi]}, S_0)] \tag{ii}$$

where ψ is $(h \geq 11)[do(move(1), now)]$

$$= \frac{1}{\gamma} \int_{x \in \mathbb{R}} \underline{\mathcal{T}[P(x, \max(0, h - 1) \geq 11, S_0)]} \tag{iii}$$

$$= \frac{1}{\gamma} \int_{x \in \mathbb{R}} P(x, \max(0, h - 1) \geq 11, S_0) \tag{iv}$$

$$= \frac{1}{\gamma} \int_{x \in \mathbb{R}} \begin{cases} p(\iota, S_0) & \text{if } \exists \iota.\, h(\iota) = x \wedge h(\iota) \geq 12 \\ 0 & \text{otherwise} \end{cases} \tag{v}$$

$$= \frac{1}{\gamma} \int_{x \in \mathbb{R}} \begin{cases} 0.1 & \text{if } x \in [2, 12] \text{ and } x \geq 12 \\ 0 & \text{otherwise} \end{cases} \tag{vi}$$

$$= \frac{1}{\gamma} \int_{x \in \mathbb{R}} \begin{cases} 0.1 & \text{if } x = 12 \\ 0 & \text{otherwise} \end{cases} \tag{vii}$$

$= 0$

We use $\mathcal{T}$'s item 5 to get (i), after which item 4(a) is applied. On doing $\mathcal{R}$ in (ii), along the lines of Example 8.5, we obtain (iii). $\mathcal{T}$'s item 4(c) then yields (iv), and stops. In the steps following (iv), we show how P expands in terms of p, and how the space of situations resolves into a mathematical expression, yielding 0.

3. $Bel(h = 0, do(move(4), S_0))$

$$= \frac{1}{\gamma} \int_{x \in \mathbb{R}} \mathcal{T}[P(x, \underline{\mathcal{R}[(h = 0)[do(move(4), now)]]}, S_0)] \qquad \text{(i)}$$

$$= \frac{1}{\gamma} \int_{x \in \mathbb{R}} \mathcal{T}[P(x, \max(0, h - 4) = 0, S_0)] \qquad \text{(ii)}$$

$$= \frac{1}{\gamma} \int_{x \in \mathbb{R}} \begin{cases} 0.1 & \text{if } x \in [2, 12] \text{ and } x \leq 4 \\ 0 & \text{otherwise} \end{cases} \qquad \text{(iii)}$$

$$= 0.2$$

By means of (6.1), after moving forward by 4 units the belief about h is characterized by a *mixed* distribution because $h = 0$ is accorded a 0.2 weight (i.e. from all points where $h \in [2, 4]$ initially), while $h \in (0, 8]$ are associated with a density of 0.1. Here, $\mathcal{T}$'s item 5 and 4(a) are triggered, and the removal of $\mathcal{T}$ using 4(c) is not shown. The end result is that the density function is integrated for $2 \leq x \leq 4$ leading to 0.2. (γ is 1.)

4. $Bel(h = 4, do(move(-4), do(move(4), S_0)))$

$$= \frac{1}{\gamma} \int_{x \in \mathbb{R}} \mathcal{T}[P(x, \exists u.\, h = u \, \wedge$$
$$4 = \max(0, u + 4), do(move(4), S_0))] \qquad \text{(i)}$$

$$= \frac{1}{\gamma} \int_{x \in \mathbb{R}} \mathcal{T}[P(x, \exists u.\, u = \max(0, h - 4) \, \wedge$$
$$4 = \max(0, u + 4), S_0)] \qquad \text{(ii)}$$

$$= \frac{1}{\gamma} \int_{x \in \mathbb{R}} \begin{cases} 0.1 & \text{if } x \in [2, 12],\ x \leq 4 \\ 0 & \text{otherwise} \end{cases} \qquad \text{(iii)}$$

$$= 0.2$$

We noted above that the point $h = 4$ gets a 0.2 weight on executing $move(4)$, after which it obtains a h value of 0. The weight is *retained* on reversing by 4 units, with the point now obtaining a h value of 4. The derivation invokes two applications of $\mathcal{T}$'s item 4(a). We skip the intermediate $\mathcal{R}$ steps. (γ evaluates to 1).

5. $Bel(h = 4, do(move(4), do(move(-4), S_0)))$

$$= \frac{1}{\gamma} \int_{x \in \mathbb{R}} \mathcal{T}[P(x, \exists u.\, u = \max(0, h + 4) \, \wedge$$
$$4 = \max(0, u - 4), S_0)] \qquad \text{(i)}$$

$$= 0$$

Had the robot moved away first, no "collapsing" of points takes place, h remains a continuous distribution and no point is accorded a non-zero weight. $\mathcal{T}$ steps are skipped but they are symmetric to the one above, e.g. compare (i) here and (ii) above. But then the density function is non-zero only for the individual $h = 4$.

6. $Bel(4 \leq h \leq 6, do(sonar(5), S_0))$

$$= \frac{1}{\gamma} \int_{x \in \mathbb{R}} \mathcal{N}(5 - x; 0, 4) \times T[P(x, \psi, S_0)] \qquad \text{(i)}$$

where ψ is $h = x \wedge 4 \leq h \leq 6$

$$= \frac{1}{\gamma} \int_{x \in \mathbb{R}} \begin{cases} 0.1 \cdot \mathcal{N}(5 - x; 0, 4) & \text{if } x \in [2, 12], \ x \in [4, 6] \\ 0 & \text{otherwise} \end{cases}$$

≈ 0.41

We obtain (i) after T's item 5 and then 4(b) for sensing actions. That is, belief about $h \in [4, 6]$ is sharpened after observing 5. Basically, we are integrating a function that is 0 everywhere except when $4 \leq x \leq 6$ where it is $0.1 \times \mathcal{N}(5 - x; 0, 4)$, normalized over $2 \leq x \leq 12$.

7. $Bel(4 \leq h \leq 6, do(sonar(5), do(sonar(5), S_0)))$

$$= \frac{1}{\gamma} \int_{x \in \mathbb{R}} \mathcal{N}(5 - x; 0, 4) \times \underline{T[P(x, \psi, s))]} \qquad \text{(i)}$$

where $s = do(sonar(5), S_0)$, ψ is $h = x \wedge 4 \leq h \leq 6$

$$= \frac{1}{\gamma} \int_{x \in \mathbb{R}} [\mathcal{N}(5 - x; 0, 4)]^2 \times T[P(x, \psi, S_0)] \qquad \text{(ii)}$$

≈ 0.52

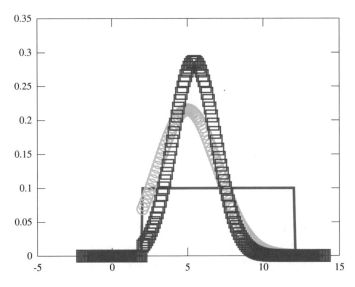

Fig. 8.1 Belief density change for h at S_0 (in blue), after sensing 5 (in green) and after sensing 5 twice (in red)

As expected, two successive observations of 5 sharpens belief further. Derivations (i) and (ii) follow from T's item 5, and two successive applications of item 4(b). Thus, we are to integrate $0.1 \times [\mathcal{N}(5 - x; 0, 4)]^2$ between [4, 6] and normalize over [2, 12]. These changing densities are plotted in Fig. 8.1.

8.3   Two Special Cases

Regression is a general property for computing properties about posteriors in terms of priors after actions. It is therefore possible to explore limited cases, which might be appropriate for some applications. We present two such cases.

Conjugate Distributions

Certain types of systems, such as Gaussian processes, admit an effective propagation model. The same advantages can be observed in our framework. We illustrate this using an example. Assume a fluent f, and suppose $\mathcal{D}_0$ is the union of P3, (P1) and the following specification:

$$p(\iota, S_0) = \mathcal{N}(f(\iota); \mu_1, \sigma_1{}^2)$$

which stipulates that the true value of f is believed to be normally distributed. Assume the following sensor in $\mathcal{D}$:

$$l(sense(z), s) = \mathcal{N}(z - f(s); \mu_2, \sigma_2{}^2)$$

Then it is easy to show that estimating posteriors yields a product of Gaussian density function (that is also a Gaussian density function), which is inferred by T:

$$T[Bel(b \leq f \leq c, do(sense(z), S_0))] =$$
$$\tfrac{1}{\gamma} \int_b^c \mathcal{N}(x; \mu_1, \sigma_1{}^2) \cdot \mathcal{N}(z - x; \mu_2, \sigma_2{}^2) dx$$

Distribution Transformations

Certain actions affect priors in a characteristically simple manner, and regression would account for these changes as an appropriate function of the initial belief state. We illustrate two instances using Example 8.8. First, consider an action $grasp(z)$ that grabs object z. Because the action of grasping does not affect h by way of (6.1), we get:

$$T[Bel(h \leq b, do(grasp(obj5), S_0))] = Bel(h \leq b, S_0)$$

So no changes to h's density are required. Second, consider ground actions with the property that two distinct values of f do not become the same after that action, e.g., for initial states this means:

$$\forall \iota, \iota'.\ f(\iota) \neq f(\iota') \supset f(do(a, \iota)) \neq f(do(a, \iota'))$$

Think of $move(-4)$ that agrees with this, but $move(4)$ need not. We can show that such actions "shift" priors:

$$\mathcal{T}[Bel(h \leq b, do(move(-n), S_0))] = Bel(h \leq b - n, S_0)$$

Intuitively, the probability of h being in the interval $[b, c]$, irrespective of the distribution family, is the same as the probability of $h \in [b + n, c + n]$ after $move(-n)$. Thus, regression derives the initial interval given the current one.

8.4   Regression over Noisy Actions

The regression operator thus far was limited to noise-free actions and noisy sensing. We first show how the logical account is extended to handle noisy actions. We then extend the regression operator.

The idea behind noisy actions is that an agent might attempt a physical move of 3 units, but as a result of the limited accuracy of effectors, actually move (say) 3.094 units. Thus, unlike sensors, where the reading is nondeterministic, *observable,* but does not affect fluents, the outcome of noisy actions is nondeterministic, *unobservable* and changes fluent properties. Of course, when attempting to move 3 units, the agent knows that an actual move by 3.094 units is much more likely than 30.94 units, and that is reflected in a noise model.

Recall that a simple approach to noisy actions was given in a previous chapter. The idea was to difference intended values and actual values in the action arguments. That is, instead of an action like $move(x)$, we introduced an action $move(x, y)$ where x is the intended motion (3 units) known to the agent and y is the actual motion (say, 3.094 units) unknown to the agent. The usual successor state axiom for h would be written, for example, as follows in the equational style:

$$h(do(a, s)) = \ \text{IF}\ \exists x, y(a = move(x, y)) \tag{8.8}$$
$$\text{THEN}\ \max(0, h(s) - y)\ \text{ELSE}\ h(s).$$

Unlike noise-free actions where the likelihood is always 1, noisy actions would have non-trivial l-axioms and might further constrain unintended outcomes:

$$l(move(x, y), s) = \text{IF}\ y = x\ \text{THEN}\ 0.9\ \text{ELSE}\ (\text{IF}\ y = 0\ \text{THEN}\ 0.1\ \text{ELSE}\ 0). \tag{8.9}$$

Together with the precondition axiom that disallows zero and negative values for x, the l-axiom says that the likelihood of y being exactly x is 0.9, that the likelihood of no move happening is 0.1 and all other possibilities are improbable. Analogously, a Gaussian noise model might be defined as follows:

$$l(move(x, y), s) = \mathcal{N}(y; x, 1).$$

This axiom says that the actual value moved is normally distributed around the intended value, with a variance of 1.

In general, we assume that for each noisy action $act(x, y)$, the action theory $\mathcal{D}$ includes an axiom of the form[2]:

$$l(act(x, y), s) = Err_{act}(x, y).$$

We are now ready to introduce term and formula regression, now with noisy actions:

Definition 8.9 We define $\mathcal{T}$ and $\mathcal{R}$ as in Definition 8.6 except with the following change to item 4(b):

If s is of the form $do(a, s')$ and a is a sensing action $sense(z)$ such that $l(sense(z), s) = Err(z, f_i(s))$ is in $\mathcal{D}$, then:
$$\mathcal{T}[P(\vec{x}, \phi, s)] = Err(z, x_i) \times \mathcal{T}[P(\vec{x}, \psi, s')]$$
where ψ is $Poss(a, now) \supset \phi \wedge f_i(now) = x_i$.
If s is of the form $do(a, s')$ and a is a noisy action $act(n, m)$ such that $l(act(u, v), s) = Err(u, v)$ is in $\mathcal{D}$ then:
$$\mathcal{T}[P(\vec{x}, \phi, s)] = \int_v Err(n, v) \times \mathcal{T}[P(\vec{x} \cdot v, \psi(v), s')]$$
where $\psi(v)$ is $Poss(act(n, v), now) \supset \mathcal{R}[\phi[do(act(n, v), now)]]$.

So, while $\mathcal{T}$ for noisy sensing is the same as before, when we have a ground noisy action $act(n, m)$, we observe the following. First, since the agent does not actually observe m, a fresh variable v is introduced, which is to be the integration variable, and the density is multiplied by the noise model. The density expression then regresses the formula w.r.t. $act(n, v)$. In fact, as we show below, this regression is best realized by keeping v as a free variable, because that then gives us a single expression over which we can integrate.

Most significantly, the desired property still holds:

Theorem 8.10 *Let $\mathcal{D}$ be any action theory, ϕ any situation-suppressed formula and δ any action sequence. Then*

$$\mathcal{D} \models Bel(\phi, do(\delta, S_0)) = \mathcal{T}[Bel(\phi, do(\delta, S_0))]$$

where $\mathcal{T}[Bel(\phi, do(\delta, S_0))]$ is a term about S_0 only.

Example 8.11 Let us consider the same domain as in Example 8.8, but in using the noisy action model from (8.8) and (8.9). Then we get the following entailments from $\mathcal{D}$:

---

[2] For ease of presentation, we are assuming that noisy actions only come with two arguments, one standing for the intended value and the other for the actual outcome. It is straightforward to generalize the likelihood axiom for k-ary actions, or even handle context-dependent noisy actions like we have enabled for noisy sensing. This is possible by allowing fluents as additional arguments in Err_{act}.

1. $Bel(h = 3 \lor h = 4, S_0) = 0,$

   $Bel(4 \leq h \leq 6, S_0) = 0.2$

   $\mathcal{T}$ does not change terms about S_0. Here, for example, the second belief term equals $\int_4^6 0.1 dx = 0.2$.

2. $Bel(h \geq 11, do(move(1, 0), S_0))$

$$= \frac{1}{\gamma} \int_{x \in \mathbb{R}} \mathcal{T}[\, P(x, \underline{h \geq 11, do(move(1, 0), S_0))}\,] \tag{i}$$

$$= \frac{1}{\gamma} \int_{x \in \mathbb{R}} \int_{v \in \mathbb{R}} Err(1, v) \times \mathcal{T}[P(x \cdot v, \underline{\mathcal{R}[\psi(v)]}, S_0)] \tag{ii}$$

   where $\psi(v)$ is $(h \geq 11)[do(move(1, v), now)]$ and $Err(1, v)$ is the error function from (8.9)

$$= \frac{1}{\gamma} \int_{x \in \mathbb{R}} \int_{v \in \mathbb{R}} Err(1, v) \times \underline{\mathcal{T}[P(x \cdot v, \max(0, h - v) \geq 11, S_0)]} \tag{iii}$$

$$= \frac{1}{\gamma} \int_{x \in \mathbb{R}} \int_{v \in \mathbb{R}} Err(1, v) \times P(x \cdot v, \max(0, h - v) \geq 11, S_0) \tag{iv}$$

$$= \frac{1}{\gamma} \int_{x \in \mathbb{R}} \int_{v \in \mathbb{R}} \begin{cases} p(\iota, S_0) \times Err(1, v) & \text{if } \exists \iota.\, h(\iota) = x \land h(\iota) \geq 11 + v \\ 0 & \text{otherwise} \end{cases} \tag{v}$$

$$= \frac{1}{\gamma} \int_{x \in \mathbb{R}} \int_{v \in \mathbb{R}} \begin{cases} 0.1 \times Err(1, v) & \text{if } x \in [2, 12] \text{ and } x \geq 11 + v \\ & \text{otherwise} \end{cases} \tag{vi}$$

$$= \frac{1}{\gamma} \int_{x \in \mathbb{R}} \int_{v \in \mathbb{R}} \begin{cases} 0.1 \times 0.9 & \text{if } x \in [2, 12] \text{ and } x \geq 12 \\ \times 0.1 & \text{if } x \in [2, 12] \text{ and } x \geq 11 \\ & \text{otherwise} \end{cases} \tag{vii}$$

$$= 0.01$$

The application of $\mathcal{T}$ and $\mathcal{R}$ proceeds in the same fashion as in Example 8.8, with the following notable changes. In (ii), $\mathcal{T}$ for noisy actions is applied, which introduces a new integration symbol (over v), and the formula to be regressed has v as a free variable. Moreover, the noise model is multiplied to the current density. As the successive steps show, regressing the formula over actions by using the RHS of successor state axioms continues to keep v as a free variable, until (vi). The noise model is such that v takes the value 1 with a probability of 0.9, and a value 0 with a probability of 0.1. We know from Example 8.8 that when the move happens by 1 unit, the regressed formula is essentially $h \geq 12$, which has a probability of 0. So the only non-zero event is when no move happens, in which case we are interested in the prior belief of $h \geq 11$, which is 0.1, but further multiplied by the likelihood of that outcome, which is also 0.1, and so we obtain 0.01.

8.5 Progression

In the worst case, regressed formulas are exponentially long in the length of the action sequence, and so it has been argued that for long-lived agents like robots, continually updating the current view of the state of the world, is perhaps better suited. Lin and Reiter proposed a theory of progression for the classical situation calculus. What we are after is an account of progression for probabilistic beliefs in the presence of stochastic noise. However, subtleties arise with the p fluent even in simple continuous domains. Recall the standard successor state axiom for h in the robot example, where an action moves the robot towards the wall but stops when the robot hits the wall:

$$h(do(a, s)) = \text{IF } \exists z(a = move(z))$$
$$\text{THEN } \max(0, h(s) - z) \text{ ELSE } h(s).$$

If the robot were to begin with (6.2) and perform the action $move(4)$, beliefs about the new value of h become much more complex. Roughly, those points where $h \in [2, 4]$ initially are mapped to a single point $h = 0$ that should then obtain a probability *mass* of 0.2, while the other points retain their initial *density* of 0.1. In effect, a probability density on h is transformed into a mixed density / distribution (and the (P3) assumption no longer holds). In the previous sections, we dealt with this issue using *regression*: beliefs are regressed to the initial state where (P3) does hold, and all the actual belief calculations can be done in the initial state.

In this section, we develop a logical theory of progression for basic action theories where such mixed distributions do not arise.[3] We provide a simple definition to support this, and then discuss general syntactic restrictions that satisfy this requirement.

---

[3] Its worth remarking that there is nothing inherently problematic about mixed distributions as far as the definability of progressed database is concerned. Indeed, for the above example, contrast the initial theory with the one after $move(4)$ below:

$$p(\iota, s) = \begin{cases} 0.1 & \text{if } h(\iota) \in [2, 12] \\ 0 & \text{otherwise} \end{cases} \qquad \text{versus} \qquad p(\iota, s) = \begin{cases} 0.2 & \text{if } h(\iota) = 0 \\ 0.1 & \text{if } h(\iota) \in (0, 8] \\ 0 & \text{otherwise} \end{cases}$$

Both of these are well-defined p-specifications. Be that as it may, it is not immediately obvious what the account of progression should look like, in terms of the general syntactic rules that allow us to update the database so as to yield the latter initial theory. There is also an issue with the definition of *Bel*, because we now need to sum over the values of h that are discrete, and integrate over the rest. Although this can be handled, it makes the account slightly more involved. The approach in this chapter eschews these complications in a simple yet reasonable manner, and moreover, as we shall shortly see, subsumes some of the analytical cases seen in the literature.

8.5.1 Invertible Action Theories

Our formulation of progression rests on introducing a new class of basic action theories, called *invertible action theories*. Recall that successor state axioms are of the general form: $f(do(a, s)) = E_f(a)[s]$, which tells us how the value of f changes from s to $do(a, s)$. We will now base our work on the following question: given the value of f at $do(a, s)$, what is the value of f at s?

Definition 8.12 Given a basic action theory $\mathcal{D}$, a fluent f is said to be *invertible* if there is an expression $H_f(a)$ uniform in *now* such that $\mathcal{D} \models f(s) = H_f(a)[do(a, s)]$. We say that $\mathcal{D}$ is invertible if every fluent in the theory is invertible.

Intuitively, a fluent is invertible when we can find a dual formulation of its successor state axiom, that is, where we can characterize the *predecessor* value of a fluent in terms of its current value.

There are three syntactic conditions on a basic action theory $\mathcal{D}$ that are sufficient to guarantee its invertibility:

i. There is an ordering on fluents such that all the fluents that appear in $E_f(a)$ other than f are earlier in the ordering.
ii. Any situation term in $E_f(a)$ appears as an argument to one of the fluents.
iii. The mapping from the value of $f(s)$ to the value of $f(do(a, s))$ given by $E_f(a)$ is bijective. (This is understood in the usual set-theoretic sense.)

Before considering some examples, here is the result:

Theorem 8.13 *If a basic action theory satisfies (i), (ii) and (iii) above, then it is invertible.*

Example 8.14 Let us consider the setting from Fig. 6.1. Recall the successor state axiom (6.5) for v. This says that $up(z)$ is the only action affecting v, thereby incorporating Reiter's monotonic solution to the frame problem. We would now equivalently write this as:

$$v(do(a, s)) = \text{IF } \exists z(a = up(z))$$
$$\text{THEN } v(s) + z \text{ ELSE } v(s).$$

This trivially satisfies (i) and (ii). The mapping from $v(s)$ to $v(do(a, s))$ is bijective and so (iii) is satisfied also. (In general, any $E_f(a)$ that is restricted to addition or multiplication by constants will satisfy (iii).) So the fluent is invertible and we have $v(s) = H_v(a)[do(a, s)]$, where $H_v(a)$ is IF $\exists z(a = up(z))$ THEN $v - z$ ELSE v.

Example 8.15 Consider (6.1). Here the mapping is not bijective because of the *max* function and the fluent *h* is not invertible. If $h(do(\alpha, s)) = 0$ where $\alpha = move(4)$, then the value of $h(s)$ cannot be determined and can be anything less than 4.

Example 8.16 Consider a successor state axiom like this:

$$v(do(a, s)) = \text{IF } \exists z(a = up(z)) \text{ THEN } (v(s))^z \text{ ELSE } v(s).$$

For $\alpha = up(2)$, we obtain a squaring function, which is not bijective. Indeed, from $v(do(\alpha, s)) = 9$, one cannot determine whether $v(s)$ was -3 or 3, and the fluent is not invertible.

Example 8.17 Consider this successor state axiom for compound interest, where v denotes the accumulated value, *rate* denotes the annual interest rate, and *lapse(z)* denotes the number of years the interest was allowed to accumulate:

$$v(do(a, s)) = \text{IF } \exists z(a = lapse(z)) \wedge relief(s) = 0$$
$$\text{THEN } v(s) \cdot (1 + rate(s))^z \text{ ELSE } v(s).$$

Suppose further:

$$rate(do(a, s)) = \text{IF } \exists z(a = change(z)) \text{ THEN } z + rate(s) \text{ ELSE } rate(s).$$

$$relief(do(a, s)) = \text{IF } a = toggle \text{ THEN } 1 - relief(s) \text{ ELSE } relief(s).$$

Here, the interest rate influences the accumulated value over the lapsed time, and *relief* being true stops the accumulation of interest. This theory is invertible, and $H_v(a)$ is given by

$$H_v(a) = \text{IF } \exists z(a = lapse(z)) \wedge relief = 0$$
$$\text{THEN } v/(1 + rate)^z \text{ ELSE } v.$$

That is, because *lapse(z)* does not affect *relief* and *rate*, we simply invert the successor state axiom for v and relativize everything to $do(a, s)$. If (say) the action *lapse(z)* also affected *rate*, by the ordering in (i), we would first obtain the H-expression for *rate* and use it in the H-expression for v.

Note that the bijection property does not prevent us from using non-bijective functions, such as squares, in the successor state axiom of v, provided that these only apply to the other fluents. (The remaining fluents essentially behave as constants at any given situation.)

Before concluding our development of invertible theories, let us reflect on the sufficiency conditions (i), (ii) and (iii). It should be clear that (iii), in fact, is also a necessary condition.

Theorem 8.18 *If a fluent is invertible then (iii) must hold.*

What about the necessity of (i) and (ii)? The main reason we insisted on these conditions in the first place is because obtaining the value of $f(s)$ from $f(do(a, s))$ becomes straightforward by using the fluent values at the start of the order. To see why dropping these conditions makes the setting challenging, suppose f and g are the only two fluents in a basic action theory, *act* is the only action, and suppose we have the following successor state axioms:

$$f(do(a, s)) = \text{ IF } (a = act)$$
$$\text{THEN } expr_1(f(s), g(s)) \text{ ELSE } f(s).$$

$$g(do(a, s)) = \text{ IF } (a = act)$$
$$\text{THEN } expr_2(f(s), g(s)) \text{ ELSE } g(s).$$

Here, $expr_i$ could denote sums or products, or any other 2-ary bijective function. Suppose we are now given the values of $f(do(a, s))$ and $g(do(a, s))$, and as motivated earlier, we are to recover the values of $f(s)$ and $g(s)$. To obtain the value of $f(s)$, then, we would need the value of $f(do(a, s))$, which is given, but also the value of $g(s)$, which has to be obtained. To obtain the value of $g(s)$, we would need the value of $g(do(a, s))$, which is given, but also the value of $f(s)$. In other words, we are given a system of equations:

$$x' = expr_1(x, y)$$
$$y' = expr_2(x, y)$$

where the values of $\{x', y'\}$ are known and denote the values of $\{f(do(a, s)), g(do(a, s))\}$ respectively, and we are to solve for $\{x, y\}$ that denote the values of $\{f(s), g(s)\}$ respectively. In many cases, such systems can, of course, be solved; for example, if $x' = x + y$ and $y' = y - x$, then $x = x' - (y' + x) = x' - y' - x$, which means $2x = x' - y'$. Once we obtain the value of x, we can obtain the value of y analogously. In general, however, solving such a system of equations may not always be possible, at least in an exact manner.

Thus, (i) and (ii) are not necessary in order to obtain H-expressions, but simplify the treatment considerably since we only need to invert the function expressed in $E_f(a)$, and obtain the values of the fluents according to the order. It would be interesting to see whether for the basic action theories considered in the literature, even if (i) and (ii) do not hold, $H_f(a)$ can be obtained in an exact manner making (iii) both sufficient and necessary for this class of theories.

8.5.2 Classical Progression

We now are prepared for a definition of progression that applies to any invertible basic action theory. Note that the definition of invertibility imposes no constraint on $\mathcal{D}_0$. So the definition in this section is general in that the p may appear in $\mathcal{D}_0$ in an unrestricted way, such as the

p-axioms leading to multiple distributions discussed in previous chapters. Given such a theory $\mathcal{D}_0 \cup \Sigma$ and a ground action α, we define a transformation $\mathcal{D}_0'$ such that $\mathcal{D}_0' \cup \Sigma$ agrees with $\mathcal{D}_0 \cup \Sigma$ on the future of α. Then, in the next section, we will consider how $\mathcal{D}_0'$ grows as a result of this progression.

To start with, let us first consider the simpler case of progression for a $\mathcal{D}_0$ that does not mention the p fluent (and the quantification over initial situations that comes with it), and so where $\mathcal{D}_0$ is uniform in S_0. In this case, because we are assuming a finite set of nullary fluents, any basic action theory can be shown to be *local-effect*, where progression is first-order definable. The new theory is computed by appealing to the notion of *forgetting*, introduced by Lin and Reiter. If the basic action theory is invertible, however, the progression can *also* be defined in another way. Let $\mathcal{D}_0'$ be $\mathcal{D}_0$ but with any $f(S_0)$ term in it replaced by $H_f(\alpha)[S_0]$.

Theorem 8.19 *Let $\mathcal{D}_0 \cup \Sigma$ be any invertible basic action theory not mentioning p and α any ground action. Then for any situation-suppressed formula ϕ:*

$$\mathcal{D}_0 \cup \Sigma \models \phi[do(\alpha, S_0)] \ \ iff \ \ \mathcal{D}_0' \cup \Sigma \models \phi[S_0].$$

Example 8.20 Consider (6.5), and the $H_v(a)$ from Example 8.14. Suppose $\mathcal{D}_0 = \{v(S_0) > 10\}$. Then:

$$\begin{aligned}
\mathcal{D}_0' &= (H_v(up(3))[S_0]) > 10 \\
&= (\text{IF } \exists z(up(z) = up(3)) \\
&\qquad \text{THEN } v(S_0) - z \text{ ELSE } v(S_0)) > 10 \\
&= (v(S_0) - 3 > 10) \\
&= (v(S_0) > 13).
\end{aligned}$$

Therefore, as expected, the progression of $v(S_0) > 10$ w.r.t. a noise-free motion of 3 units is $v(S_0) > 13$. (The unique name axiom and arithmetic are used in the simplification.)

8.5.3 Progressing Degrees of Belief

There are two main complications when progressing beliefs w.r.t. noisy sensors and actions. First, the p fluent will have to take the likelihood of the action α into account. Second, $\mathcal{D}_0$ need not be uniform in S_0, since p typically requires quantification over initial situations (as in (6.2), for example). This leads to the following definition:

Definition 8.21 Let $\mathcal{D}_0 \cup \Sigma$ be an invertible basic action theory and α be a ground action of the form $A(\vec{t})$ where $\vec{t}$ is uniform in now.[4] Then $Pro(\mathcal{D}_0, \alpha)$ is defined as $\mathcal{D}_0$ with the following substitutions:

---

[4] In the most common case (like noise-free or sensing actions), the arguments to the action would simply be a vector of constants.

- $p(\iota, S_0)$ is replaced by $\dfrac{p(\iota, S_0)}{Err_A(\vec{t})[\iota]}$;
- every other fluent term $f(u)$ is replaced by $H_f(\alpha)[u]$.

Here, $Err_A(\vec{x})$ refers to the RHS of the likelihood axiom for $A(\vec{x})$. The intuition of having the likelihood function in the denominator is this: consider that $p(\iota, S_0) = \phi(\iota)$ determines the weight of the initial situation ι. After doing an action, we would want to adjust the weight of that situation depending on the action performed and the value observed on the sensor, and whether that value is in agreement with the fluent measured in ι. The weight adjustment is realized using the likelihood function $Err(a)$: that is, $p(\iota, S_0) = \phi(\iota) \times Err(a)$; in other words, $p(\iota, S_0)/Err(a) = \phi(\iota)$.

The main result is the correctness of this definition of progression:

Theorem 8.22 *Under the conditions of the definition above, let $\mathcal{D}_0' = Pro(\mathcal{D}_0, \alpha)$. Suppose that $\mathcal{D}_0 \models (Err_A(\vec{t}) \neq 0)[S_0]$. Then for any situation-suppressed formula ϕ:*

$$\mathcal{D}_0 \cup \Sigma \models \phi[do(\alpha, S_0)] \;\; iff \;\; \mathcal{D}_0' \cup \Sigma \models \phi[S_0].$$

This theorem gives us the desired property for *Bel* (which is defined in terms of p) as a corollary:

Corollary 8.23 *Suppose $\mathcal{D}_0$, Σ, $\mathcal{D}_0'$, ϕ, and α are as above. Then for all real numbers n:*

$$\mathcal{D}_0 \cup \Sigma \models Bel(\phi, do(\alpha, S_0)) = n$$
$$iff \quad \mathcal{D}_0' \cup \Sigma \models Bel(\phi, S_0) = n.$$

Thus the degree of belief in ϕ after a physical or sensing action is equal to the initial belief in ϕ in the progressed theory.

We now present some examples, considering, in turn, noise-free actions, noisy sensing and finally noisy actions.

Example 8.24 Let us consider an action theory with a vertical action $up(z)$, a sensing action $sonar(z)$ and two horizontal actions: *towards* moves the robot halfway towards the wall and *away* moves the robot halfway away from the wall. Formally, let $\mathcal{D}_0 \cup \Sigma$ be an action theory where $\mathcal{D}_0$ contains just (6.2), and Σ includes

- foundational axioms and (P1)-P3 as usual;
- a l-axiom for $sonar(z)$, namely (8.7);
- l-axioms for the other actions, which are noise-free, and so these simply equal 1;
- a successor state axiom for v, namely (6.5);

- the following successor state axiom for h:

$$h(do(a, s)) =$$
$$\text{IF } a = away \text{ THEN } 3/2 \cdot h(s)$$
$$\text{ELSE IF } a = towards \text{ THEN } 1/2 \cdot h(s)$$
$$\text{ELSE } h(s).$$

We noted that (6.1) does not satisfy our invertibility property. This variant, however, is invertible. The H-expression for v was derived in Example 8.14. The H-expression for h is:

$$H_h(a) =$$
$$\text{IF } a = away \text{ THEN } 2/3 \cdot h$$
$$\text{ELSE IF } a = towards \text{ THEN } 2 \cdot h$$
$$\text{ELSE } h.$$

We now consider the progression of $\mathcal{D}_0$ w.r.t. the action $away$. First, the instantiated H-expressions would simplify to:

- $H_h(away) = 2/3 \cdot h$;
- $H_v(away) = v$.

Next, since $away$ is noise-free, we have $Err_{away} = 1$. Putting this together, we obtain $\mathcal{D}'_0 = Pro(\mathcal{D}_0, away)$ as:

$$p(s, S_0) = \mathcal{U}(2/3 \cdot h; 2, 12) \times \mathcal{N}(v; 0, 1) \, [s]$$
$$= \mathcal{U}(h; 3, 18) \times \mathcal{N}(v; 0, 1) \, [s]$$

That is, the new p is one where h is uniformly distributed on [3, 18] and v is independently drawn from a standard normal distribution (as before). This leads to a *shorter* and *wider* density function, as depicted in Fig. 8.2. Here are three simple properties to contrast the original versus the progressed:

- $\mathcal{D}_0 \cup \Sigma \models Bel(h \geq 9, S_0) = 0.3$.
  The *Bel* term expands as:

$$\frac{1}{\gamma} \int_x \int_y \begin{array}{l} \text{IF } \exists \iota (h = x \wedge v = y \wedge h \geq 9)[\iota] \\ \text{THEN } \mathcal{U}(h; 2, 12) \times \mathcal{N}(v; 0, 1)[\iota] \ \text{ELSE } 0 \end{array}$$

which simplifies to the integration of a density function:

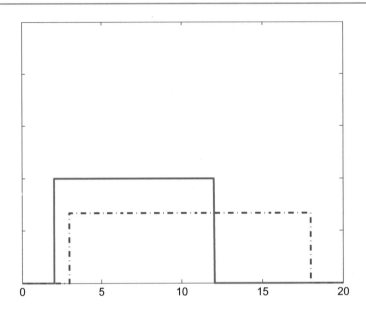

Fig. 8.2 Belief change about h: initially (solid magenta) and after moving away (dotted blue)

$$\frac{1}{\gamma} \int_x \int_y \begin{cases} 0.1 \times \mathcal{N}(y; 0, 1) & \text{if } x \in [2, 12], x \geq 9 \\ 0 & \text{otherwise} \end{cases}$$

$$= \frac{1}{\gamma} \int_x \int_y \begin{cases} 0.1 \times \mathcal{N}(y; 0, 1) & \text{if } x \in [9, 12] \\ 0 & \text{otherwise} \end{cases}$$

$$= 0.3.$$

Only those situations where $h \in [2, 12]$ initially are given non-zero p values and by the formula in the *Bel*-term, only those where $h \geq 9$ are to be considered.

- $\mathcal{D}_0 \cup \Sigma \models Bel(h \geq 9, do(away, S_0)) = 0.6$.
  For any initial situation ι, $h[do(away, \iota)] \geq 9$ only when $h[\iota] \geq 6$, which is given an initial belief of 0.6.
- $\mathcal{D}'_0 \cup \Sigma \models Bel(h \geq 9, S_0) = 0.6$.
  Basically, *Bel* simplifies to an expression of the form:

$$\frac{1}{\gamma} \int_x \int_y \begin{cases} 1/15 \cdot \mathcal{N}(y; 0, 1) & \text{if } x \in [3, 18], x \geq 9 \\ 0 & \text{otherwise} \end{cases}$$

giving us 0.6.

Example 8.25 Let $\mathcal{D}_0 \cup \Sigma$ be exactly as above, and consider its progression w.r.t. *towards*. It is easy to verify that for instantiated H-expressions we get:

- $H_h(\textit{towards}) = 2 \cdot h$;
- $H_v(\textit{towards}) = v$;

Here too, because *towards* is noise-free, $Err_{towards}$ is simply 1, which is to say the $\mathcal{D}_0' = Pro(\mathcal{D}_0, \textit{towards})$ is defined as:

$$p(s, S_0) = \mathcal{U}(2 \times h; 2, 12) \times \mathcal{N}(v; 0, 1) \, [s]$$
$$= \mathcal{U}(h; 1, 6) \times \mathcal{N}(v; 0, 1) \, [s].$$

The new distribution on h is *narrower* and *taller*, as shown in Fig. 8.3. Here we might contrast $\mathcal{D}_0$ and $\mathcal{D}_0'$ as follows:

- $\mathcal{D}_0 \cup \Sigma \models Bel(h \in [2, 3], S_0) = 0.1$.
- $\mathcal{D}_0' \cup \Sigma \models Bel(h \in [2, 3], S_0) = 0.2$.

Example 8.26 Let $\mathcal{D}_0 \cup \Sigma$ be as in the previous examples. Consider its progression w.r.t. the action *sonar*(5). Sensing actions do not affect fluents, so for H-expressions we have:

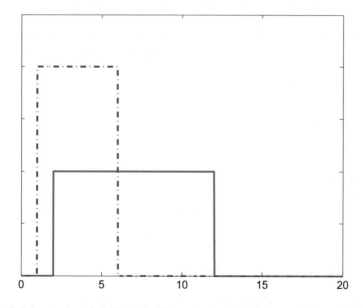

Fig. 8.3 Belief change about h: initially (solid magenta) and moving towards the wall (dotted blue)

- $H_h(sonar(5)) = h$;
- $H_v(sonar(5)) = v$.

Here $sonar(z)$ is noisy, and we have $Err_{sonar}(5) = \mathcal{N}(5; h, 4)$. This means that the progression $\mathcal{D}'_0 = Pro(\mathcal{D}_0, sonar(5))$ is

$$\frac{p(\iota, S_0)}{\mathcal{N}(5; h, 4)[\iota]} = \mathcal{U}(h; 2, 12) \times \mathcal{N}(v; 0, 1)[\iota],$$

which simplifies to the following:

$$p(\iota, S_0) = \mathcal{U}(h; 2, 12) \times \mathcal{N}(v; 0, 1) \times \mathcal{N}(5; h, 4) \, [\iota].$$

As can be noted in Fig. 8.4, the robot's belief about h's true value around 5 has sharpened. Consider, for example, that:

- $\mathcal{D}_0 \cup \Sigma \models Bel(h \leq 9, S_0) = 0.7$.
- $\mathcal{D}'_0 \cup \Sigma \models Bel(h \leq 9, S_0) \approx 0.97$.

If we were to progress $\mathcal{D}'_0$ further w.r.t. a second sensing action, say $sonar(5.9)$, we would obtain the following:

$$p(\iota, S_0) = \\ \mathcal{U}(h; 2, 12) \times \mathcal{N}(v; 0, 1) \times \mathcal{N}(5; h, 4) \times \mathcal{N}(5.9; h, 4) \, [\iota].$$

As can be seen in Fig. 8.4, the robot's belief about h would sharpen significantly after this second sensing action. If we let $\mathcal{D}''_0 = Pro(\mathcal{D}'_0, sonar(5.9))$ then:

- $\mathcal{D}''_0 \cup \Sigma \models Bel(h \leq 9, S_0) \approx 0.99$.

Example 8.27 Let $\mathcal{D}_0$ be (6.2). Let Σ be the union of:

- (P1)-P3 and domain-independent foundational axioms;
- a successor state axiom for h as above;
- a noisy move action up with the following l-axiom:

$$l(up(x, y), s) = \mathcal{N}(y; x, 2)$$

- a successor state axiom for v using this noisy move:

$$v(do(a, s)) = \text{ IF } \exists x, y(a = up(x, y)) \\ \text{THEN } v(s) + y \text{ ELSE } v(s).$$

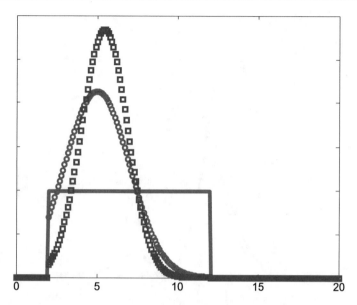

Fig. 8.4 Belief change about h: initially (solid magenta), after sensing 5 (red circles), and after sensing twice (blue squares)

(Recall that for a noisy move $up(x, y)$, x is the intended motion and y is the actual motion.) This is inverted using the same idea as in Example 8.14.

Consider the progression of $\mathcal{D}_0 \cup \Sigma$ w.r.t. $up(2, 3)$, where, of course, the agent does not get to observe the latter argument, and so corresponds to the action $up(2, u)$. The simplified H-expressions are as follows:

- $H_h(up(2, u)) = h$;
- $H_v(up(2, u)) = v - u$.

By definition, occurrences of v in $\mathcal{D}_0$ are to be replaced by $H_v(up(2, u))$. Also, $Err_{up}(2, u) = \mathcal{N}(u; 2, 2)$. Therefore, $\mathcal{D}_0' = Pro(\mathcal{D}_0, up(2, u))$ is defined to be

$$\left(\frac{p(\iota, S_0)}{\mathcal{N}(u; 2, 2)[\iota]} = \mathcal{U}(h; 2, 12) \times \mathcal{N}(v - u; 0, 1)[\iota] \right)$$

This simplifies to:

$$p(\iota, S_0) = \mathcal{U}(h; 2, 12) \times \mathcal{N}(v; u, 1) \times \mathcal{N}(u; 2, 2) \ [\iota].$$

Thus the noisy action has had the effect that the belief about the position has shifted by an amount u drawn from a normal distribution centered around 2. This leads to a *shifted* and

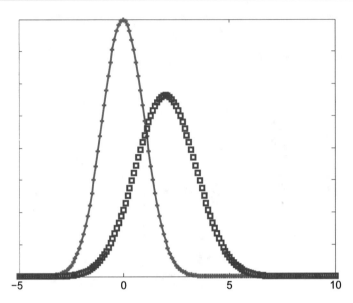

Fig. 8.5 Belief change about v: initially (solid magenta) and after a noisy move of 2 units (blue squares)

wider curve seen in Fig. 8.5. As expected, the agent is considerably less confident about its position after a noisy move. Here, for example, are the degrees of belief about being located within 1 unit of the best estimate (that is, the mean):

- $\mathcal{D}_0 \cup \Sigma \models Bel(v \in [-1, 1], S_0) \approx 0.68$.
- $\mathcal{D}'_0 \cup \Sigma \models Bel(v \in [1, 3], S_0) \approx 0.34$.

Basically, *Bel* expands to an expression of the form

$$\frac{1}{\gamma} \int_{x,y,z} \begin{cases} 0.1\mathcal{N}(y; z, 1) \cdot \mathcal{N}(z; 2, 2) & \text{if } x \in [2, 12], y \in [1, 3] \\ & \text{otherwise} \end{cases}$$

where γ is

$$\int_{x,y,z} \begin{cases} 0.1\mathcal{N}(y; z, 1) \cdot \mathcal{N}(z; 2, 2) & \text{if } x \in [2, 12] \\ & \text{otherwise} \end{cases}$$

leading to 0.34.

8.6   Computability of Progression

In the general case, the computability of progression is a major concern, as it requires second-order logic. We are treating a special case here, and because it is defined over simple syntactic transformations, we have the following result immediately:

Theorem 8.28 *Suppose $\mathcal{D} = \mathcal{D}_0 \cup \Sigma$ is any invertible basic action theory. After the iterative progression of $\mathcal{D}_0 \cup \Sigma$ w.r.t. a sequence δ, the size of the new initial theory is $O(|\mathcal{D}| \times |\delta|)$.*

Therefore, progression is computable. But for realistic robotic applications, even this may not be enough, especially over millions of actions. Consider, for example, that to calculate a degree of belief it will be necessary to integrate the numerical expression for p. What we turn to in this section is a special case that would guarantee that over any length of action sequences, the size of the progressed theory does not change beyond a constant factor. It will use the notion of *context-completeness* (as considered by Liu and Levesque) and a few simplification rules.

Definition 8.29 Suppose $F \subseteq \{f_1, \ldots, f_k\}$ is any set of fluents, and $\mathcal{D}_0 \cup \Sigma$ is any invertible basic action theory. We say that $\mathcal{D}_0$ is *complete* w.r.t. F if for any $\phi \in Lang(F)$, either $\mathcal{D}_0 \models \phi$ or $\mathcal{D}_0 \models \neg\phi$, where $Lang(F)$ is the sublanguage restricted to the fluents in F.

Definition 8.30 An invertible basic action theory $\mathcal{D}_0 \cup \Sigma$ is said to be *context-complete* iff

- for every fluent f, $\mathcal{D}_0$ is complete w.r.t. every fluent other than f appearing in the successor state axiom of f;
- $\mathcal{D}_0$ is complete w.r.t. every fluent appearing in a conditional expression in the likelihood axioms.

That is, there is sufficient information in $\mathcal{D}_0$ to simplify all the conditionals appearing in the context formulas of the successor state axioms and the likelihood axioms.

STRIPS actions are trivially context-complete, and so are so-called *context-free* successor state axioms where only rigid symbols appear in the RHS.[5] In Example 8.17, if $\mathcal{D}_0$ is complete w.r.t. the fluents *rate* and *relief*, then the theory would be context-complete. Note that $\mathcal{D}_0$ does not need to be complete w.r.t. the fluent v in that example, and this is precisely why they are interesting. Indeed, both (6.1) and (6.5) are also context-complete because, by definition, E_f may mention f, and (say) use its previous value. The reader may further

---

[5] The background theory in STRIPS is a database; preconditions are simply facts that must be true in the database for the action to be executable. Actions affect change to the database in terms of additions and deletions of facts.

verify that all the density change examples developed here are context-complete. Since we are interested in iterated progression, we say that the progression of $\mathcal{D}$ over an action sequence $\delta = [a_1, \ldots, a_k]$ is context-complete iff *the iterated progression is context-complete*: that is, $\mathcal{D}_0 \cup \Sigma$ is context-complete, $Pro(\mathcal{D}_0, a_1) \cup \Sigma$ is context-complete, $Pro(Pro(\mathcal{D}_0, a_1), a_2) \cup \Sigma$, and so on.

Putting it all together, in contrast to our previous theorem, we would have a progressed theory that in linear in the size of the initial theory $\mathcal{D}_0$:

Theorem 8.31 *Suppose $\mathcal{D}_0 \cup \Sigma$ is any invertible basic action theory that is also context-complete. After the iterative progression of $\mathcal{D}_0 \cup \Sigma$ w.r.t. a sequence δ, the size of the new initial theory is $O(|\mathcal{D}_0| + |\delta|)$.*

This can make a substantial difference in the size of the expression for p. A special case of this theorem is immediately applicable to *conjugate distributions*, previously considered in Sect. 8.3. Indeed, such distributions admit an effective propagation model, as seen in Kalman filtering. We show a simple example where analogous expressions are obtained by our definition of progression:

Example 8.32 Let $\mathcal{D}_0 \cup \Sigma$ be as in Example 8.27. We noted its progression w.r.t. $up(2, 3)$ includes:

$$p(\iota, S_0) = \mathcal{U}(h; 2, 12) \times \mathcal{N}(v; z, 1) \times \mathcal{N}(z; 2, 2) [\iota].$$

If we progress this sentence further w.r.t. a second noisy action $up(3, 4)$, we would obtain:

$$p(\iota, S_0) = \mathcal{U}(h; 2, 12) \times \mathcal{N}(v - u; y, 1) \times \mathcal{N}(y; 2, 2) \times \mathcal{N}(u; 3, 2) [\iota].$$

This then simplifies to:

$$p(\iota, S_0) = \mathcal{U}(h; 2, 12) \times \mathcal{N}(v; y + u, 1) \times \mathcal{N}(y; 2, 2) \times \mathcal{N}(u; 3, 2) [\iota].$$

8.7 Notes

Planning and robotic applications have to deal with numerous sources of complexity regarding action and change. Along with efforts in related knowledge representation formalisms such as dynamic logic [227], the action language [87] and the fluent calculus [218], Reiter's [190] reconsideration of the situation calculus has proven enormously useful for the design of logical agents, essentially paving the way for *cognitive robotics* [138].

In this work, we obtained new results on how to handle projection in the presence of probabilistic information, both at the level of the knowledge base and at the level of actions. In particular, we generalized both regression and progression [190]. Its worth noting that classical (that is, non-probabilistic) regression and progression have also been investigated for other formalisms. For example, [65] study regression in a variant of dynamic logic, and fluent calculus is based on a type of progression [214, 216].

Our regression results are interesting because irrespective of the decompositions and factorizations that are justifiable initially, belief state evolution is known to invalidate these factorizations even over simple temporal phenomena [36]. (See [59, 98] for other instances of such problems.) We demonstrated regression in settings where actions affect priors in nonstandard ways, such as transforming a continuous distribution to a mixed one. In general, regression does not insist on (but allows[6]) restrictions to actions, that is, no assumptions need to be made about how actions affect variables and their dependencies over time. Moreover, at the specification level, we do not assume (but allow) structurally constrained initial states.

Given the generality of our results, and the promising advances made in the area of relational probabilistic inference [29, 58, 58, 94, 184], regression suggests natural ways to apply those developments with actions. This line of research would allow us to address effective belief propagation for numerous planning problems that require both logical and probabilistic representations. On another front, note that after applying the reductions, one may also use approximate inference methods. Perhaps then, regression can serve as a computational framework to study approximate belief propagation, on the one hand, and using approximate inference at the initial state after goal regression, on the other. In [26], an implementation of the regression operator is considered. For the sake of concreteness, it allows any single probability distribution, including factorized representations such as Bayesian networks, which is evaluated by sampling. It should be possible to repurpose the action reduction with any other p-specification.

With regards to progression, Lin and Reiter developed their notion with long-lived agents in mind [158]. (The argument on regression leading to formulas exponential in the action sequence was given in [190].) Lin and Reiter also showed that progression is always second-order definable, and in general, it appears that second-order logic is unavoidable [229]. Since then, a number of other special cases have been studied, such as [160], which showed that progression with *local-effect* action theories was first-order definable. This result used the notion of forgetting, which was introduced in [157]. Context-completeness was considered in [161],[7] and context-free successor state axioms (including STRIPS actions) are discussed in Reiter's book [190].

However, their account did not deal with probabilistic uncertainty nor with noise, as seen in real-world robotic applications. In the work here, we consider semantically correct

---

[6] In particular, the tractability of Gaussian propagation models and conjugate distributions are discussed in [30, 35, 219].

[7] Strictly speaking, our notion of context-completeness is inspired by, but not the same as the one in [161]. This pertains to the fact that we allow successor state axioms to use the fluent's previous value.

progression in the presence of continuity.[8] By first identifying what we called invertible basic action theories, we obtained a new way of computing progression.[9] Under the additional restriction of context-completeness, progression is very efficient. Most significantly, by working within a richer language, we have obtained progression machinery that, to the best of our knowledge, has not been discussed elsewhere, and goes beyond existing techniques. The unrestricted nature of the specification of the p fluent, for example, which we inherit from [8], allows for agents whose beliefs are not determined by a unique distribution. There are two immediate directions for future work on progression. First, just like regression was implemented in [26], it would be worthwhile to investigate an implementation for progression. Second, the invertibility property was mainly sought to handle continuity, including the case where a continuous distribution transforms to a discrete one. If we restrict our attention to discrete distributions, the natural question is whether one can obtain an account of progression in stochastic domains that does not syntactically restrict the basic action theory.

Finally, building on both our results, developing a planner that leverages the ideas behind regression and progression, as one would in classical planning [85, 92], would make for very interesting future work. In less expressive formalisms, probabilistic planners based on a type of belief regression have been in development [119].

---

[8] We deviate from the formulation in [158] in one minor way: the classical account defines the progression of $\mathcal{D}_0$ to be a set of sentences that are uniform in $do(\alpha, S_0)$, and require that $\mathcal{D}_0$ and $\mathcal{D}_0'$ agree on all formulas about $do(\alpha, S_0)$ and its future situations. (Note that ϕ in the theorem is permitted to mention now, which means it can capture future situations by being of the form $\psi[do([\alpha_1, \ldots, \alpha_n], now)]$.) So, the theorem would instead be stated as saying that $\mathcal{D}_0 \cup \Sigma \models \phi[do(\alpha, S_0)]$ iff $\mathcal{D}_0' \cup \Sigma \models \phi[do(\alpha, S_0)]$. We are choosing to instead formulate $\mathcal{D}_0'$ as being uniform in S_0, as we think its somewhat simpler to read when we invert the successor state axioms.

The original variant is easily obtained by simply replacing all occurrences of S_0 in $\mathcal{D}_0'$ by $do(\alpha, S_0)$. Replacing the situation terms in a progressed theory is not uncommon; see, for example, the lifting of predicate symbols in [160].

[9] For discussions on invertible functions in real analysis, and the inspiration for our bijective construction $E_f(a)$, see [222].

Programs

9

> *If the world were merely seductive, that would be easy. If it were*
> *merely challenging, that would be no problem. But I arise in the*
> *morning torn between a desire to improve the world and a desire to*
> *enjoy the world. This makes it hard to plan the day.*
>
> —*E. B. White*

High-level program execution, as seen in GOLOG and FLUX among others, offers an attractive alternative to automated planning for the control of artificial agents. This is especially true for the dynamic worlds with incomplete information seen in robotics, where planning would have to generate complex behavior that includes branches and loops. One of research goals of the area of *cognitive robotics* is to understand what sorts of high-level behavior specifications would be both useful and tractable in settings like these.

A major criticism leveled at much of this work, however, is that it makes unrealistic assumptions about the robots themselves, whose sensors and effectors are invariably noisy and best characterized by continuous probability distributions. When a robot is executing a GOLOG program involving some condition ϕ about the world, for example, it is unreasonable to expect that it will either know ϕ to be true, know it to be false, or know nothing about it. To remedy this, in previous chapters, we discussed a logical account for reasoning about *probabilistic degrees of belief* in the presence of *noise*.

In this chapter, we propose a new *implemented programming formalism* called ALLEGRO that is intended as an alternative to GOLOG for high-level control in robotic applications.

9.1 From Knowledge-Based to Belief-Based Programs

To appreciate the need for this extension, recall that knowledge-based programs enabled us to condition programs on what is known and what is observed. For our task of getting the robot close to the wall, we might have:

© The Author(s), under exclusive license to Springer Nature Switzerland AG 2023   149
V. Belle, *Toward Robots That Reason: Logic, Probability & Causal Laws*,
Synthesis Lectures on Artificial Intelligence and Machine Learning,
https://doi.org/10.1007/978-3-031-21003-7_9

until *Knows(Close)* **do** *senseact* **endUntil**

where *senseact* stands for the sequence: (*sonar*; *safeact*). Furthermore, *safeact* is moving only when the robot is not close:

if *Knows(¬Close)* **then** *move* **endIf**

Let us consider how such a program should be written in the presence of noise. Based on what we have discussed previous chapters, the following difficulties can be seen to arise:

1. Sensing is no longer a binary entity, so how would we incorporate the value read on the sensor?
2. In the *senseact* program, after performing the first sonar action, the degree of belief in *h* increases but is not certain, so what should *safeact* look like?
3. Repeated sensing clearly further increases the agent's belief about *h*'s value, so how many sensing actions are appropriate?
4. In the *safeact* program, the forward action brings the robot closer but also introduces noise, so how should we account for the changing degrees of belief about *h* in the program?

Intuitively, we address these difficulties in the following manner:

- Regarding (1) above, recall that previous chapters modeled projection queries such that the value read was an argument to the ground action. Clearly this is not sensible when writing programs as the modeler cannot be expected to know what sort of values might get read. We will show how we can use a nullary action for sensing but nonetheless have the values read be used during program execution.
- Regarding (2), sensing enables the agent to have posterior beliefs about the distribution of *h*'s values. We will introduce the notion of *degree of confidence,* defined in terms of the *p*-fluent like *Bel*, but where we can better articulate that the "spread" of probabilities on *h*'s values is over a narrower range, much like in Fig. 8.4.
- Regarding (3), as we cannot know in advance how many sensing actions might be required for the belief about *h* to be narrow, we can use the **until** construct to repeatedly perform sensing. The argument to this construct can involve a formula with the confidence operator.
- For (4), analogous to our sensing actions, we will continue using physical actions with only the intended argument, so it is not unlike noise-free GOLOG programs. The actual outcome nonetheless gets used during during program execution.
- To better relate the noise of effectors with the increase in confidence obtained by sensing, we will need to define sense-act loops. Basically, the robot has gotten closer to the wall but is less confident about *h*'s values owing to the noise introduced on acting.

As with standard GOLOG, the user provides a *basic action theory* (or BAT) to describe the domain in terms of fluents, actions and sensors, possibly characterized by discrete and continuous distributions. Given such a BAT, ALLEGRO allows the user to specify iterative programs that appeal to the robot's beliefs, and is equipped with an efficient methodology for program execution.[1] Overall, the proposal incorporates the following aspects:

- ALLEGRO can be used for projection, but it is not just a projection system. It interprets standard GOLOG constructs (e.g., while-loops) while adapting GOLOG's primitive programs for noise and outcome unknowns. An implementation-independent semantic characterization of programs can be provided.
- ALLEGRO can be used in offline (i.e., knowing sensed outcomes in advance), *online* (i.e., sense as the robot is acting in an unknown environment) and *network* (e.g., controlling a robot over TCP) modes.
- To handle iterative programs involving noisy actions, ALLEGRO is realized in terms of an efficient interpreter based on *sampling* and *progression*. Inspired by particle filtering systems, the interpreter maintains a database of sampled states corresponding to the agent's initial uncertainty and updates these as the program executes. For one thing, this allows belief-level queries in programs to be evaluated efficiently in an online setting. For another, sampling admits a simple strategy for handling the nondeterminism in noisy actions. Nonetheless, the interpreter is argued to be correct using limits. To our knowledge, no other GOLOG implementation embodies such techniques, and such a correctness notion is new to the high-level programming literature.

We will now introduce the language, and then discuss its properties.

9.2   The ALLEGRO System

The ALLEGRO system is a programming language with a simple LISP-like syntax.[2] In this section, we first informally discuss how domain axiomatizations are written in ALLEGRO, formally introduce the grammar of ALLEGRO programs, and discuss how ALLEGRO is used.

---

[1] Note that the specification of belief-based programs in the situation calculus can be given at a more general level than considered in this chapter. Our implementation of the ALLEGRO system, for example, commits to a single probability distribution, but the program semantics does not hinge on this technical limitation, and the full language can be supported. Our view was that it would be worthwhile to entertain the design and implementation of one possible practical instance of the general language, and this chapter is as good as any for that purpose.

[2] The ALLEGRO system is realized in the RACKET dialect of the SCHEME family (racket-lang.org). We use RACKET arithmetic freely, such as the max function, as well as any other function that can be defined in RACKET, like GAUSSIAN. However, the technical development does not hinge on any feature unique to that language.

9.2.1 Domain Axiomatization

As a small example, Fig. 9.1 shows a BAT for our robot example from Fig. 3.2. To help the user follow the syntax, we recap how the axiomatization is to be read:

1. The domain has a single fluent h, the distance to the wall, whose initial value is taken by the agent to be drawn from a (continuous) uniform distribution on [2, 12].
2. The successor state axiom for h says that it is affected only by the move action. The first argument of move is the amount the agent intends to move, and the second is the amount that is actually moved (which determines how much the value of h is to be reduced).
3. The alt axiom is used to say that if an action of the form (move 2 2.79) occurs, the agent will only know that (move 2 z) happened for some value of z.
4. The likelihood axiom for move says that the actual amount moved by move will be expected by the agent to be centered (w.r.t. a normal distribution) around the intended amount with a standard deviation of 1.0.
5. The likelihood axiom for sonar says that sonar readings are noisy but will be expected to be centered around the true value of h with a standard deviation of 4.0.

The idea here is that starting with some initial beliefs, executing (sonar 5) does not guarantee that the agent is actually 5 units from the wall, although it should serve to increase the robot's confidence in that fact. Analogously, performing a noisy move with an intended argument 2 means that the robot may end up moving (say) 2.79 units. Nevertheless, its degree of belief that it is closer to the wall should increase.

```
(define-fluents h)
(define-ini-p-expr '(UNIFORM h 2 12))
(define-ss-exprs h
   (move x y) '(max 0 (- h ,y))
(define-alts
   (move x y) (lambda (z) '(move ,x ,z)))
(define-l-exprs
   (move x y) '(GAUSSIAN ,y ,x 1.0)
   (sonar z) '(GAUSSIAN ,z h 4.0))
```

Fig. 9.1 A BAT for the simple robot domain

9.2.2 Belief-Based Programs

The BAT specification describes the noise in actions and sensors w.r.t. the actual outcomes and values observed. But a robot executing a program need not know these outcomes and values. For this reason, the *primitive programs* of ALLEGRO are actions that suppress these parameters. So for the actions (move x y) and (sonar z) appearing in a BAT, the primitive programs will be (move x) and (sonar).

The basic ALLEGRO language uses five program constructs:

prim	primitive programs;
(begin $prog_1$ ... $prog_n$)	sequence;
(if *form* $prog_1$ $prog_2$)	conditional;
(let ((var_1 $term_1$) ...(var_n $term_n$)) *prog*)	assignments;
(until *form* *prog*)	until loop.

Here *form* stands for formulas built from this grammar:

$$form:: = (\circ\ term_1\ term_2)\ |\ (\bullet\ form_1\ form_2)\ |\ (\text{not}\ form)$$

where $\circ \in \{<, >, =\}$ and $\bullet \in \{\text{and}, \text{or}\}$. Here *term* stands for terms built from the following grammar:

$$term:: = (\text{exp}\ term)\ |\ number\ |\ fluent\ |\ var\ |\ (\diamond\ term_1\ term_2)\ |$$
$$(\text{if}\ form\ term_1\ term_2)$$

where $\diamond$ is any arithmetic operator (e.g., + and −). The primary "epistemic" operator in ALLEGRO is exp: (exp *term*) refers to the *expected value* of *term*. (Reasoning about the expected value about a fluent allows the robot to monitor how it changes with sensing, for example.) The *degree of belief* in a formula *form* can then be defined in terms of exp as follows:

$$(\text{bel}\ form) \doteq (\text{exp}\ (\text{if}\ form\ 1.0\ 0.0))$$

For our purposes, it is convenient to also introduce (conf *term number*), standing for the *degree of confidence* in the value of a term, as an abbreviation for:

$$(\text{bel}\ (>\ number\ (\text{abs}\ (-term(exp\ term)))))$$

For example, given a fluent f that is normally distributed, (conf f 0.1) is higher when the curve is narrower.

9.2.3 Usage

The ALLEGRO system allows programs to be used in three ways: in *online* mode—the system displays each primitive program as it occurs, and prompts the user to enter the sensing results; in *network* mode—the system is connected to a robot over TCP, the system sends primitive programs to the robot for execution, and the robot sends back the sensing data it obtains; and finally, in *offline* mode—the system generates ersatz sensing data according to the given error model. In all cases, the system begins in an initial belief state, and updates this belief state as it runs the program.

As a simple illustration, imagine the robot from Fig. 9.1 would like to get within 2 and 6 units from the wall. It might proceed as follows: sharpen belief about current position (by sensing), (intend to) move by an appropriate amount, adjust beliefs for the noise of the move, and repeat these steps until the goal is achieved. This intuition is given as a program in Fig. 9.2: here, conf is used to first become confident about h and then exp is used to determine the distance for getting midway between 2 and 6 units. (An arbitrary threshold of 0.8 is used everywhere w.r.t. the robot's beliefs.) For an online execution of this program prog in ALLEGRO, we would do:

```
> (online-do prog)
Execute action: (sonar)
Enter sensed value: 4.1
Enter sensed value: 3.4
Execute action: (move 1.0)
Enter sensed value: 3.9
Enter sensed value: 4.2
Execute action: (move 0.0)
```

We see the robot first applying the sonar sensor, for which the user reports a value of 4.1. After updating its beliefs for this observation, the robot is not as confident as required by prog, and so, a second sensing action is executed for which 3.4 is read. Then the robot attempts a (noisy) move, but its confidence degrades as a result. So these steps are repeated once more, after which the program terminates. On termination, the required property can be tested in ALLEGRO using:

```
(until (> (bel (and (>= h 2) (<= h 6))) .8)
   (until (> (conf h .4) .8) (sonar))
   (let ((diff (- (exp h) 4)))
     (move diff)))
```

Fig. 9.2 A program to get between 2 and 6 units from the wall

```
> (bel (and (>= h 2) (<= h 6)))
0.8094620133032484
```

Analogously, the offline mode follows the execution style except that the user does not provide the observed values. As explained, the system generates ersatz sensing data according to the given error model, so we might see an execution as follows:

```
> (offline-do prog)
Action received: (sonar)
Returning sensor value of 4.9711099203570654
Action received: (sonar)
Returning sensor value of 6.941606856002471
Execute action: (move 1.939588384369502)
...
```

This would continue until the robot is within 2 and 6 units from the wall, analogous to the execution for the online mode.

With the TCP mode, the values are provided over TCP. It is initiated by running the following on the first terminal:

```
racket -if examples/prog.scm
```

This leads to:

```
> (prog)
Connecting to action server on TCP port 8123
```

Then, on the second terminal, we run the server:

```
racket -fm action-server.scm

Waiting for localhost TCP connection on port 8123
Client has connected
Action received: (sonar)
Returning sensor value of 6.9
Action received: (sonar)
...
```

Here too, we would provide the observed value, after which the execution proceeds analogously until the robot is within the required distance from the wall. As can be surmised, the semantics of program execution does not fundamentally depend on the exact mode that the program is run in; in contrast to the online mode, the offline and TCP modes simply provide a different means to return observed values to the program. So let us now turn to how the semantics can be defined formally.

9.3 Mathematical Foundations

The formal foundation of ALLEGRO, not surprisingly, is based on the situation calculus. In particular, the BAT syntax of ALLEGRO can be seen as a situation-suppressed fragment of the situation calculus language.[3]

Example 9.1 For the sake of concreteness, note that the sonar's noise model, move action's nondeterminism and initial beliefs of the robot from Fig. 9.1 can be mapped to the following situation calculus formulas: $l(sonar(z), s) = Gaussian(z; h, 4)[s]$, $alt(move(x, y), z) = move(x, z)$ and $p(s, S_0) = Uniform(h; 2, 12)[s]$.

Recall that the *degree of belief* in any situation-suppressed ϕ at s is defined using:

$$Bel(\phi, s) \doteq \frac{1}{\gamma} \int_{\vec{x}} P(\vec{x}, \phi, s)$$

where the normalization factor γ is obtained by replacing ϕ with *true*. Given an argument ϕ, the P-term returns the p-value of the situation if ϕ holds, and 0 otherwise. By extension, the *expected value* of a situation-suppressed term t at s can be defined as:

$$Exp(t, s) \doteq \frac{1}{\gamma} \int_{\vec{x}} (t \times P(\vec{x}, true, now))[s]$$

where the denominator γ replaces t with 1. This is to be read as considering the t-values across situations and multiplying them by the p-values of the corresponding situations. So, if a space of situations is uniformly distributed, then we would obtain the average t-value.

9.3.1 Program Semantics

ALLEGRO implements GOLOG over online executions and belief operators. Recall from Sect. 3.2.6 that GOLOG programs are defined over the following constructs:

$$\alpha \mid \phi? \mid (\delta_1; \delta_2) \mid (\delta_1 \mid \delta_2) \mid (\pi\ x)\delta(x) \mid \delta^*.$$

---

[3] Non-logical symbols, such as fluent and action symbols, in the situation calculus are italicized; e.g., fluent h in ALLEGRO is h in the language of the situation calculus.

Programming constructs (used in ALLEGRO), such as if-then-else, while and until loops, can then be defined in terms of these, as shown in Sect. 3.2.6.

There are a few notable differences to standard GOLOG, however. For one thing, ϕ would mention *Bel* or *Exp*, analogous to knowledge-based GOLOG programs using *Knows*. For another, and perhaps more significantly, the smallest programs in our setting are not atomic actions, but new symbols called *primitive programs* (Sect. 9.2.2), which serve as placeholders for actual actions on execution. So for every $a(\vec{x}, y)$ in the logical language, where y corresponds to the actual outcome/observation, there are primitive programs $a(\vec{x})$ that suppress the latter argument.

A semantics for the online execution of ALLEGRO programs over noisy acting and sensing can be given in the very same style as for classical GOLOG. The full details are not necessary for our discussion, and we only sketch the main idea. Two special predicates *Trans* and *Final* are introduced to define a single-step transition semantics for programs: $Trans(\delta, s, \delta', s')$ means that by executing program δ at s, one can get to s' with an elementary step with the program δ' remaining, whereas $Final(\delta, s)$ holds when δ can successfully terminate in situation s. The predicates *Trans* and *Final* are defined axiomatically. For example, the transition requirements for sequence are:

$$Trans([\delta_1; \delta_2], s, \delta', s') \equiv$$
$$Final(\delta_1, s) \wedge Trans(\delta_2, s, \delta', s') \vee$$
$$\exists \delta''[Trans(\delta_1, s, \delta'', s') \wedge \delta' = (\delta''; \delta_2)]$$

which says that to single-step $(\delta_1; \delta_2)$, either δ_1 terminates and we single-step δ_2, or we single-step δ_1 leaving some program δ'' and so $(\delta''; \delta_2)$ is what is left. The only difference in these definitions between classical GOLOG and ALLEGRO is the handling of noisy actions and sensors in the latter. If actions are noise-free, the transition requirements are as follows:

$$Trans(a(\vec{x}), s, \delta', s') \equiv$$
$$Poss(a(\vec{x}), s) \wedge \delta' = nil \wedge s' = do(a(\vec{x}), s).$$

This says that the noise-free action is executable, and on doing it to reach the successor state, only the empty program is left.

For ALLEGRO, recall that noisy actions and sensing are missing an argument. Therefore, if $a(\vec{x})$ is a primitive program in ALLEGRO, and on doing this action we obtain a value c (either an observation or the outcome of a noisy action) then:

$$Trans(a(\vec{x}), s, \delta', s') \equiv$$
$$Poss(a(\vec{x}, c), s) \wedge \delta' = nil \wedge s' = do(a(\vec{x}, c), s).$$

These axiomatic definitions are lumped as a set $\mathcal{E}$, along with an encoding of programs as first-order terms for quantificational purposes as a set $\mathcal{F}$. Putting it all together, we say that the *online execution* of a program δ *successfully terminates* after a sequence of actions σ if

$$\mathcal{D} \cup \mathcal{E} \cup \mathcal{F} \models Do(\delta, S_0, do(\sigma, S_0))$$

where $Do(\delta, s, s') \doteq \exists \delta' \, [Trans^*(\delta, s, \delta', s') \land Final(\delta', s')]$ and $Trans^*$ denotes the reflexive transitive closure of $Trans$.

9.4   A Sampling-Based Interpreter

We now present pseudo-code for the interpreter of ALLEGRO and argue for its correctness relative to the specification above. The interpreter is based on sampling, and its correctness is defined using limits.

The overall system is described using three definitions: an evaluator of expressions EVAL, an epistemic state e, and an interpreter of programs INT. In what follows, we let a *world state* w be a vector of fluent values (i.e., $w[i]$ is the value of the ith fluent) and an *epistemic state* e is a finite set of pairs (w, q) where w is a world state and $q \in [0, 1]$. Intuitively, w is a world considered possible by the agent, and q is the probabilistic weight attached to that world.

The Evaluator

The interpreter relies on an evaluator EVAL that takes three arguments: a term or a formula, a world state, an epistemic state, and returns a value (corresponding to the value of the term or the truth value of a formula):

Definition 9.2 EVAL$[\cdot, w, e]$ is defined inductively over a term t or formula d as follows. For terms, the operator returns a number or a variable. For formulas, the operator returns 0 or 1.

1. EVAL$[u, w, e] = u$, when u is a number or a variable.
2. EVAL$[f, w, e] = w[i]$, when f is the ith fluent.
3. EVAL$[(\text{not } d), w, e] = 1 - \text{EVAL}[d, w, e]$.
4. EVAL$[(\text{and } d_1 \, d_2), w, e] = 1$ iff EVAL$[d_1, w, e] = \text{EVAL}[d_2, w, e] = 1$, and similarly for other logical connectives over formulas.
5. EVAL$[(+ \, t_1 \, t_2), w, e] = \text{EVAL}[t_1, w, e] + \text{EVAL}[t_2, w, e]$, and similarly for other arithmetic operators over terms.
6. EVAL$[(\text{if } d \, t_1 \, t_0), w, e] = \text{EVAL}[t_i, w, e]$, where we obtain $i = \text{EVAL}[d, w, e]$.
7. EVAL$[(\text{exp } t), w, e] = \left. \sum_{(w', q) \in e} \text{EVAL}[t, w', e] \times q \, \middle/ \, \sum_{(w', q) \in e} q \right.$

We write EVAL$[t, e]$ when w is not used (because all the fluents in t appear in the scope of an exp), and EVAL$[t, w]$ when e is not used (because t contains no exp terms).

Epistemic State

To relate the evaluator to a BAT specification, we first obtain the initial epistemic state e_0. To get the idea, suppose the values of the fluents f_i were to range over finite sets $dom(f_i)$. Given any $\mathcal{D}_0$ of the form $\forall s (p(s, S_0) = \Theta[s])$, we let:[4]

$$e_0 = \{(w, q) \mid w[i] \in dom(f_i) \text{ and } q = \text{EVAL}[\Theta, w]\}.$$

By construction, e_0 is the set of all possible worlds in a finite domain setting. However, in continuous domains, $dom(f_i)$ is $\mathbb{R}$, leading to infinitely many possible worlds. Nonetheless, this space can be approximated in terms of a finite epistemic state by sampling: assume we can randomly draw elements (and vectors) of $\mathbb{R}$. Suppose the BAT is defined using k fluents, and let n be a large number. Then:

$$e_0 = \{(w_i, q_i) \mid \text{choose } w_1, \ldots, w_n \in \mathbb{R}^k, \text{ and } q_i = \text{EVAL}[\Theta, w_i]\}.$$

When actions occur, the interpreter below also depends on a *progression* procedure PROG that takes an epistemic state and an action instance as arguments and returns a new state:

Definition 9.3 Suppose $a(c)$ is a noisy sensor. Then

$$\begin{aligned}
\text{PROG}[e, a(c)] = \{ (w', q') \mid\ & (w, q) \in e, \\
& w'[j] = \text{EVAL}[\text{SSA}_{f_j}(a(c)), w], \\
& q' = q \times \text{EVAL}[\text{LH}_a(c), w] \}
\end{aligned}$$

where SSA and LH refer to the RHS of successor state and likelihood axioms in $\mathcal{D}$ instantiated for $a(c)$.

Intuitively, for $(w, q) \in e$, standing for some current situation and its density, we use the given BAT to compute the fluent values and the density of the successor situation. In particular, for a noisy sensor, the successor density incorporates the likelihood of the observed value c.

Definition 9.4 Suppose $a(c, c')$ is a noisy action. Then

$$\begin{aligned}
\text{PROG}[e, a(c, c')] = \{ (w'_i, q'_i) \mid\ & \text{choose } d_1, \ldots, d_n \in \mathbb{R}, \\
& (w_i, q_i) \in e, \\
& w'_i[j] = \text{EVAL}[\text{SSA}_{f_j}(a(c, d_i)), w_i], \\
& q'_i = q_i \times \text{EVAL}[\text{LH}_a(c, d_i), w_i] \}.
\end{aligned}$$

---

[4] We can evaluate the RHS of axioms in $\mathcal{D}$ w.r.t. a world state using Definition 9.2. These are simply formulas (with a single situation term) of the logical language, and in the ALLEGRO syntax, they correspond to expressions mentioning the fluents. For example, evaluating the function (UNIFORM h 2 12) at a world state where h is 10 would yield 0.1, but in one where it is 15 would yield 0.

The notion of progression is more intricate for noisy actions because for any ground action term $a(c, c')$, the agent only knows that $a(c, y)$ happened for some $y \in \mathbb{R}$. (For ease of exposition, assume $Alt(a(c, x), y) = a(c, y)$.) Among other things, this means that the progression would need to account for all possible instantiations of the variable y, which are infinitely many. Sampling, however, permits a simple strategy: assume we can randomly draw n elements of $\mathbb{R}$, once for each world state in e, the idea being that the value drawn is applied as an argument to action a for the corresponding world. Consequently, for any e and a, $|e| = |\text{PROG}[e, a(c, c')]|$, and the sampling limit is still shown to converge. In essence, we needed to appeal to sampling for each occurrence of an integration variable in Exp: for every fluent and for the argument of every noisy action.

The Interpreter

Finally, we describe the interpreter for ALLEGRO, called INT. It takes as its arguments a program and an initial epistemic state and returns as its value an updated epistemic state:

Definition 9.5 INT$[\delta, e]$ is defined inductively over δ by

1. INT$[(a),e] = \text{PROG}[e, a(c)]$, where c is observed on doing noisy-sensor primitive program a.
2. INT$[(a\ t),e] = \text{PROG}[e, a(c, c)]$, where $c = \text{EVAL}[t, e]$, after doing noisy-action primitive program a for argument c. (Since the second argument of a is irrelevant for e, we simply set it also to c.)
3. INT$[(\text{begin}\,\delta_1 \ldots \delta_n),e] = \text{INT}[\delta_n, \text{INT}[(\text{begin}\,\delta_1 \ldots \delta_{n-1}),e]]$.
4. INT$[(\text{if}\,\phi\,\delta_1\,\delta_0), e] = \text{INT}[\delta_i, e]$, where $i = \text{EVAL}[\phi, e]$.
5. INT$[(\text{let}\,((x\ t))\,\delta), e] = \text{INT}[\delta_v^x, e]$, where $v = \text{EVAL}[t, e]$.
6. INT$[(\text{until}\,\phi\,\delta), e] = e$ if $\text{EVAL}[\phi, e] = 1$, and otherwise, INT$[(\text{until}\,\phi\,\delta), \text{INT}[\delta, e]]$.

The interpreter's correctness is established by the following:

Theorem 9.6 *Suppose $\mathcal{D}$ is a BAT with e_0 (of size n) as above, δ is any program, and t is any term. Suppose there is an action sequence σ such that $\mathcal{D} \cup \mathcal{E} \cup \mathcal{F} \models Do(\delta, S_0, do(\sigma, S_0))$. Then,*

$$\mathcal{D} \models Exp(t, do(\sigma, S_0)) = u \text{ iff } \lim_{n\to\infty} \text{EVAL}[(\text{exp}\,t), \text{INT}[\delta, e_0]] = u.$$

In sum, we provide a particle-filtering type account for a set of worlds sampled from the initial distribution. The expected value of any term interpreted over that set of worlds matches the logically entailed number in the limit.

Let us reiterate the intuition here. When dealing with finite distributions and noise-free actions, an initial epistemic state can be constructed such that the expected values are precisely the entailed values. Mainly, we need to construct an epistemic state that includes every possible world that is accorded non-zero probability; this is doable because distributions are assumed to be finite. Since actions are noise-free, the successors to the possible initial worlds are unique. If actions are noisy, however, there may be many such successors to the initial worlds. And so by not increasing the cardinality of the successor epistemic state in relation to e_0, we now have an approximation. Likewise, with continuous distributions, as discussed earlier, the finite e_0 is an approximation. In the limit, then, the expected values of terms are the same numbers.

9.5   Notes

The ALLEGRO system is a programming model based on the situation calculus, introduced in [27], and so is a new addition to the GOLOG family of high-level programming languages [146, 198]. The semantics for the interpreter was based on [198]; for a discussion on sampling, see [173].

The ALLEGRO system, in fact, follows in the tradition of *knowledge-based* GOLOG with sensing in an online context [48, 74, 191], but generalizes this in the sense of handling degrees of belief and probabilistic noise in the action model.

The GOLOG family has been previously extended for probabilistic nondeterminism, but there are significant differences. For example, in the PGOLOG model [97], the outcome of a nondeterminism action is immediately *observable* after doing the action, and continuous distributions are not handled. This is also true of the notable DTGOLOG approach [34] and its variants [75]. In this sense, the ALLEGRO model is more general where the agent cannot observe the outcomes and sensors are noisy. Moreover, these proposals do not represent beliefs explicitly, and so do not include a query language for reasoning about nested belief expressions.

Outside of the situation calculus, an alternative to GOLOG, called FLUX [216, 217], has been extended for knowledge-based programming with noisy effectors in [164, 215]; continuous probability distributions and nested belief terms are, however, not handled. The concept of knowledge-based programming was first introduced in [73]. These have been extended for Spohn-style [207] ordinal functions in [142]. For discussions on how high-level programming relates to standard agent programming [204], see [200].

As discussed in previous chapters, programs, broadly speaking, generalize sequential and tree-like plan structures, but can also be used to limit plan search [9]. There are, of course, many planning approaches in online contexts, including knowledge-based planning [181, 226], decision-theoretic proposals [118, 186], corresponding relational abstractions [196, 233], and belief-based planning [120]. See [140] on viewing knowledge-based programs as plans.

In the same vein, let us reiterate that the BAT syntax is a simplified fragment of the probabilistic situation calculus introduced earlier. For ease of exposition, we limited discussions on ALLEGRO in the following ways. First, we omitted any mention of action pre-conditions. Second, fluents are assumed to be real-valued. In general, as seen in previous chapters, fluent values can range over any set, BATs are not limited to any specific family of discrete/continuous distributions, and the language supports notions not usually seen in standard probabilistic formalisms, such as *contextual* likelihood axioms. (For example, if the floor is slippery, then the noise of a move action may be amplified.) All of these are fully realized in ALLEGRO. It should be possible to also add other features, such as support for non-unique distributions, by incorporating appropriate inference machinery in ALLEGRO. This makes the language distinct from probabilistic planning languages such as [195] as the BAT syntax allows for arbitrary successor state axioms, and it rests on a very general model for logic and probability.

In the context of probabilistic models, we remark that there are other realizations of program-based probabilistic behavior, such as *probabilistic programming* [95, 169]. These are formal languages that provide program constructs for probabilistic inference. While they are expressive enough to capture dynamical probabilistic models [177], they belong to a tradition that is different from the GOLOG and FLUX families. For example, atomic programs in GOLOG are actions taken from a basic action theory whereas in [169], atomic constructs can be seen as random variables in a Bayesian Network. In other words, in GOLOG, the emphasis is on high-level control, whereas in many probabilistic programming proposals, the emphasis is on inference. So, a direct comparison is difficult; whether these traditions can be combined is an open question.

Before wrapping up, let us emphasize two issues concerning our interpreter. First, it is worth relating the interpreter's sampling semantics to the notion of progression of BATs. In a prior chapter, we investigated this latter notion of progression for continuous domains. Progression is known to come with strong negative results, and in continuous domains, we noted that further restrictions to *invertible* basic action theories may be necessary. In a nutshell, invertible theories disallow actions such as the *max* function (used in Fig. 9.1) because they have the effect of transforming a uniform continuous distribution on h to one that is no longer purely continuous. In contrast, no such limitations are needed here: roughly, this is because our notion of an epistemic state is a discrete model of a continuous function, leading to a possible world-based progression notion. Consequently, however, we needed to appeal to limits here, while the progression formulation in that prior work does not.

Second, our discussion on regression and progression so far assumed an interpreter based entirely on one of these approaches. However, a direction pursued by [71] maintains the sequence of performed actions and processes these via regression to progress effectively. (For example, the progression of a database w.r.t. an action whose effects are undone by another action in the future can be avoided.) Moreover, interpreters may appeal to regression when some form of plan search is necessary within programs [138]. The consideration of such ideas in the ALLEGRO context would lead to useful extensions. (For example, if the performed noisy actions are characterized by normal distributions, then their conjugate property can be exploited via regression or progression.)

A Modal Reconstruction

10

Show not what has been done, but what can be. How beautiful the world would be if there were a procedure for moving through labyrinths.
—**Umberto Eco**, *The Name of the Rose*

The previous chapters considered an extension to Reiter's reworking of the situation calculus to allow for probabilistic belief and noise. By introducing a functional fluent for weights on situations, and a likelihood function for outcomes, we captured belief as an abbreviation in terms of the p-values of situations.

Since the situation calculus is defined axiomatically, no special semantics is needed. However, when we wish to consider theoretical questions that are not direct entailments of basic action theories, involved arguments based on Tarskian structures or considerable proof-theoretic machinery is needed. In the non-probabilistic epistemic situation calculus alone, elementary questions about knowledge, such as: from $K\alpha \supset (K\beta \vee K\gamma)$, does it follow that $K\alpha \supset K\beta$ or $K\alpha \supset K\gamma$ require multi-page proofs. This situation is clearly much worse if we are arguing about degrees of beliefs. Moreover, when we think about modeling a partially specified domain, our intuition is that a quantitative account subsumes a qualitative specification, and an account with actions subsumes a non-dynamic specification. Put differently, in the absence of noise, the p-extension should behave just like K-extension, and in the further absence of actions, we should be left with first-order logic. No such relation has been established previously, and it is fair to surmise that deriving such a result would be very involved for the above reasons. Finally, despite being a model of belief, we did not really consider any meta-belief properties: for example, what does *introspection* look like with degrees of belief?

As discussed previously, the account of knowledge and belief in the situation calculus is inspired by modal logic, but it does not quite have the simplicity of modal logic for constructing arguments about epistemic operators. In this chapter, we reconsider the model of belief from our previous chapters, and propose a new logical variant that has much of the expres-

© The Author(s), under exclusive license to Springer Nature Switzerland AG 2023
V. Belle, *Toward Robots That Reason: Logic, Probability & Causal Laws*,
Synthesis Lectures on Artificial Intelligence and Machine Learning,
https://doi.org/10.1007/978-3-031-21003-7_10

sive power of the original. However, this is no easy task, because of the non-propositional nature of the existing proposal. In particular, recall that B was defined in terms of sums of p-values of situations, of which there are uncountably many. Analogously, we would need a way to define beliefs in the presence of uncountably many worlds. Addressing this matter, as well as things like the likelihoods of actions and observational indistinguishability in a modal logic leads to a crisper view of the underlying framework. Modal logic is also widely used and studied in philosophy and formal verification, and such a reconstruction of our framework allows for an easier and more direct comparison.

Along the way, we are able to introduce a companion modality to K.[1] Consider that $K\phi$ says that ϕ is known, and we use $\neg K\phi$ to mean that ϕ is not known. Now, $K\phi$ does not preclude $K\psi$ for some $\psi \neq \phi$. In this way, $K\phi$ is really saying that ϕ is "at least" known. It would be convenient to say something is all that is known, which would provide a succinct way to capture what is known but also what is not known. Levesque was among the first to capture this idea of *only knowing* in a modal logic. We will see how this operator can now be considered in the presence of probabilities.

We structure the chapter as follows. We begin by introducing the modal reconstruction of the non-probabilistic *epistemic situation calculus*, the logic $\mathcal{ES}$. We then introduce the probabilistic variant, the logic $\mathcal{DS} = \mathcal{ES}+$ *degrees of belief*. In both logics, the language is built so as to reason about beliefs and meta-beliefs over actions in a first-order setting. Quantification, in particular, is understood substitutionally w.r.t. a fixed countably infinite set of *rigid designators* that exist in all possible worlds.

10.1 The Non-probabilistic Case

The non-modal fragment of $\mathcal{ES}$ consists of standard first-order logic with $=$. That is, connectives $\{\wedge, \forall, \neg\}$, syntactic abbreviations $\{\exists, \equiv, \supset\}$ defined from those connectives, and a supply of variables variables $\{x, y, \ldots, u, v, \ldots\}$. Different to the standard syntax, however, is the inclusion of (countably many) *standard names* (or simply, names) for both objects and actions $\mathcal{C}$, which will allow a simple, substitutional interpretation for $\forall$ and $\exists$. These can be thought of as special extra constants that satisfy the unique name assumption and an infinitary version of domain closure.

Like in the situation calculus, to model immutable properties, we assume rigid predicates and functions, such as *IsPlant(x)* and *father(x)* respectively. To model changing properties, $\mathcal{ES}$ includes fluent predicates and functions of every arity, such as *Broken(x)* and *height(x)*. Note that there is no longer a situation term as an argument in these symbols to distinguish the fluents from the rigids. For example, $\mathcal{ES}$ also includes a distinguished fluent predicates *Poss* and to model the executability of actions and capture sensing outcomes respectively, but they are now a unary predicates. Terms and formulas are constructed as usual. The set

---

[1] Recall our (implicit) convention for epistemic operators: we use bold letters, such as K, for modal logics, and plain italics, such as *Knows*, for the classical situation calculus.

of ground atoms $\mathcal{P}$ are obtained by applying all object names in $\mathcal{C}$ to the predicates in the language.

There are three modal operators in $\mathcal{ES}$: $[a]$, $\Box$ and $\mathbf{K}$. For any formula α, we read $[a]\alpha$, $\Box\alpha$ and $\mathbf{K}\alpha$ as "α holds after a", "α holds after any sequence of actions" and "α is known," respectively. In classical situation calculus parlance, we would use $[a]\alpha$ to capture successor situations as properties that are true after an action in terms of the current state of affairs. Together with the $\Box$ modality, which allows to capture quantification over situations and histories, basic action theories can be defined. Like in the classical approach, one is interested in the entailments of the basic action theory.

10.1.1 Semantics

Recall that in the simplest setup of the possible-worlds semantics, worlds mapped propositions to $\{0, 1\}$, capturing the (current) state of affairs. $\mathcal{ES}$ is based on the very same idea, but extended to dynamical systems. So, suppose a world maps $\mathcal{P}$ and $\mathcal{Z}$ to $\{0, 1\}$.[2] Here, $\mathcal{Z}$ is the set of all finite sequences of action names, including the empty sequence $\langle\rangle$. Let $\mathcal{W}$ be the set of all worlds, and $e \subseteq \mathcal{W}$ be the *epistemic state*. By a *model*, we mean a triple (e, w, z) where $z \in \mathcal{Z}$.

Intuitively, each world can be thought of a situation calculus tree, denoting the properties true initially but also after every sequence of actions. $\mathcal{W}$ is then the set of all such trees. Given a triple (e, w, z), w denotes the real world, and z the actions executed so far. Interestingly, e captures the accessibility relation between worlds, but by modeling the relation as a set, we are enabling positive and negative introspection using a simple technical device.

To account for how knowledge changes after (noise-free) sensing, one defines $w' \sim_z w$, which is to be read as saying "w' and w agree on the sensing for z", as follows:

- if $z = \langle\rangle$, $w' \sim_z w$ for every w'; and
- $w' \sim_{z \cdot a} w$ iff $w' \sim_z w$, $w[Poss(a), z] = 1$ and $w'[SF(a), z] = w[SF(a), z]$.

This is saying that initially, we would consider all worlds compatible, but after actions, we would need the world w' to agree on the executability of actions performed so far as well as agree on sensing outcomes. The reader might notice that this is clearly a reworking of the successor state axiom for K introduced previously.

With this, we get a simply account for truth:

- $e, w, z \models p$ iff p is an atom and $w[p, z] = 1$;
- $e, w, z \models \alpha \wedge \beta$ iff $e, w, z \models \alpha$ and $e, w, z \models \beta$;
- $e, w, z \models \neg\alpha$ iff $e, w, z \not\models \alpha$;

---

[2] We need to extend the mapping to additionally interpret fluent functions and rigid symbols, omitted here for simplicity.

- $e, w, z \models \forall x \alpha$ iff $e, w, z \models \alpha_n^x$ for all $n \in \mathcal{C}$;
- $e, w, z \models [a]\alpha$ iff $e, w, z \cdot a \models \alpha$;
- $e, w, z \models \Box \alpha$ iff $e, w, z \cdot z' \models \alpha$ for all $z' \in \mathcal{Z}$;
- $e, w, z \models K\alpha$ iff for all $w' \sim_z w$, if $w' \in e$, $e, w', z \models \alpha$.

To define entailment for a logical theory, we write $\mathcal{D} \models \alpha$ to mean for every $M = (e, w, \langle\rangle)$, if $M \models \alpha'$ for all $\alpha' \in \mathcal{D}$, then $M \models \alpha$. We write α to mean $\{\} \models \alpha$.

10.1.2 Properties

Let us first begin by observing that given a model (e, w, z), we do not require $w \in e$. It is easy to show that if we stipulated the inclusion of the real world in the epistemic state, $K\alpha \supset \alpha$ would be true. That is, suppose $K\alpha$. By the definition above, w is surely compatible with itself after any z, and so α must hold at w. Analogously, properties regarding knowledge can be proven with comparatively simpler arguments in a modal framework, in relation to the classical epistemic situation calculus. Valid properties include:

1. $\Box(K(\alpha) \wedge K(\alpha \supset \beta) \supset K(\beta))$;
2. $\Box(K(\alpha) \supset K(K(\alpha)))$;
3. $\Box(\neg K(\alpha) \supset K(\neg K(\alpha)))$;
4. $\Box(\forall x.\, K(\alpha) \supset K(\forall x.\, \alpha))$; and
5. $\Box(\exists x.\, K(\alpha) \supset K(\exists x.\, \alpha))$.

Note that such properties hold over all possible action sequences, which explains the presence of the $\Box$ operator on the outside. The first is about the closure of modus ponens within the epistemic modality. The second and third are on positive and negative introspection. The last two reason about quantification outside the epistemic modality, and what that means in terms of the agent's knowledge. For example, item 5 says that if there is some individual n such that the agent knows $Teacher(n)$, it follows that the agent believes $\exists x Teacher(x)$ to be true. This may seem obvious, but note that the property is really saying that the existence of an individual in some possible world implies that such an individual exists in all accessible worlds. It is because there is a fixed domain of discourse that these properties come out true; they are referred to a the Barcan formula.

We will now revisit the litmus test example in $\mathcal{ES}$. But before explaining how successor state axioms are written, one might wonder whether a successor state axiom for K is needed, as one would for *Knows* in the epistemic situation calculus. It turns out because the compatibility of the worlds already accounted for the executability of actions and sensing outcomes in accessible worlds, such an axiom is actually a property of the logic:

$$\models \Box[a]K(\alpha) \equiv [SF(a) \wedge K(SF(a) \supset [a]\alpha)] \vee [\neg SF(a) \wedge K(\neg SF(a) \supset [a]\alpha)].$$

Thus, what will be known after an action is understood in terms of what was known previously together with the sensing outcome. The example below will further clarify how SF works.

10.1.3 Axiomatization: The One-Dimensional Robot

Let us revisit the robot example from Fig. 3.2 in $\mathcal{ES}$. We are essentially translating the basic action theory from Sect. 4.2.5. That is, suppose the *move* action is only possible when the robot is not at the wall, and moving forward reduces the distance to the wall by one. There is a single sensor *sonar* that lets the robot know if its less than 4 units away. Then, we have the following precondition, sensing and successor axioms respectively:

- $\Box Poss(a) \equiv (a = move \wedge h \neq 0) \vee (a = sonar \wedge true)$.
- $\Box SF(a) \equiv (a = move \wedge true) \vee (a = sonar \wedge h \leq 4)$.
- $\Box[a]h = x \equiv (a = move \wedge h = x + 1) \vee (a \neq move \wedge h = x)$.

Lumping these axioms as Σ, let us further stipulate that while the robot is 5 units away from the wall, but it is completely ignorant about the value of h. We might capture this situation either by looking at the entailments of:

$$\mathcal{D}_0 \wedge \Sigma \wedge \boldsymbol{K}(\Sigma) \wedge \forall x. \neg \boldsymbol{K}(h \neq x). \tag{10.1}$$

Shortly, we will introduce the semantics for an *only knowing* modal operator O, which captures all that is known by the agent. Using that operator, we might instead consider the entailments of the following sentence instead:

$$\mathcal{D}_0 \wedge \Sigma \wedge \boldsymbol{O}(\Sigma).$$

Here, $\mathcal{D}_0 = \{h = 5\}$. The former theory is the $\mathcal{ES}$-version of the theory from Sect. 4.2.5, where only *Knows* is used. The latter theory is the more succinct version where by stipulating that Σ is only known, it follows that nothing is known about h's value initially. Either way, let $Close = h \leq 4$. It is now possible to reproduce the same set of truths as in Sect. 4.2.5.

Example 10.1 The following are entailments of (10.1):

1. $\neg \boldsymbol{K} Close$.
2. $[move] \neg \boldsymbol{K} Close$.
3. $[move][sonar] \boldsymbol{K} Close$.

The fact that we were able to recreate action theories and get appropriate entailments from Sect. 4.2.5 is by design, of course, as $\mathcal{ES}$ is meant as a modal variant of the epistemic situation calculus. But the relation between these two formalisms is much stronger: it can be shown that any $\mathcal{ES}$-formula ϕ not mentioning names can be translated to a well-defined situation calculus formula ϕ' such that ϕ is valid in $\mathcal{ES}$ if and only if ϕ' is valid in the situation calculus. This means, as far as analyzing entailments go, $\mathcal{ES}$ comes without any loss of expressiveness compared to the standard situation calculus. Moreover, it is a faithful and correct reconstruction. Owing to its simple model theory based on possible worlds, we have a crisp modal language to explore logical properties. We will shortly see whether a modal language for the p-variant can be proposed too.

10.1.4 Beyond the Semantics

The modal variant affords simpler proofs for properties, as mentioned already, but there are a number of other extensions that are worth noting.

For one thing, like the classical variant, a regression operator $\mathcal{R}$ can be introduced such that formulas with the K and $[a]$ modalities can be reduced to what is known initially:

Theorem 10.2 *Suppose ϕ is a formula only mentioning the K and $[a]$ modalities. Suppose Σ is the union of the precondition, successor state and sensing axioms. Suppose $\alpha \wedge \Sigma \wedge K(\beta \wedge \Sigma) \models \alpha$. Then $\mathcal{R}[\phi]$ is a formula without action modalities such that $\alpha \wedge K\beta \models \mathcal{R}[\phi]$.*

Here, α denotes what is true in the real world initially and β what is believed initially. So, by regression, we reduce ϕ to a formula that can be reasoned from α and β alone.

A second issue that arises in this regard is the effort it takes to express everything that is known, but *also what is not known*. In the robot example, we explicitly had to declare that nothing is known about the distance to the wall. It would be convenient to model a "closure" operator that captures what is believed, such that everything outside the scope of the operator can be taken to be not known.

Interestingly, a very simple account can be given to such an operator. Let $O\alpha$ be read as "all that is known in α." Then:

- $e, w, z \models O\alpha$ iff for all $w' \sim_z w$, $w' \in e$ iff $e, w', z \models \alpha$.

Notice that the definition of truth for O differs only in the "iff" rather than the "if" for K. That is, the worlds in e are precisely those where α is true. With such an operator, it can be shown that:

$$\mathcal{D}_0 \wedge \Sigma \wedge O(\Sigma) \models \neg K(Close).$$

That is, we only need to say that the basic action theory is only known, and since nothing is specified about what is true initially from an epistemic viewpoint, we get $\neg K(Close)$.

More generally, the new operator relates to K in a manner that really captures the closure property we desired. For example:

- $O\alpha \models K\beta$ for any β such that $\alpha \models \beta$, which says that only-knowing a formula means also knowing all of its consequences; but
- $O\alpha \models \neg K\beta$ iff $\alpha \not\models \beta$, which says that the only formulas known are the consequences of α.

But the convenience is only part of the gain. Recall that regression already reduced formulas to what is known initially. If were to use only knowing, there is a further reduction to pure first-order reasoning:

Theorem 10.3 *Suppose α and β are formulas without modalities, and ϕ a formula only mentioning the K modality. Then there is a first-order formula ψ (constructed from ϕ and β) such that:*

$$\alpha \wedge O\beta \models \phi \text{ iff } \alpha \models \psi.$$

It is finally worth noting that any valid $\mathcal{ES}$-formula, as long as it does not mention standard names and O, can be mapped onto a valid epistemic situation calculus formula. So, $\mathcal{ES}$ is really just a modal reworking of the classical situation calculus, providing all the same features but in a much simpler formal framework.

10.2 Allowing Probabilities

The situation calculus with the p-fluent accorded weights to situations, modeled nondeterministic outcomes using alt, and captured the likelihood of these outcomes using the fluent l. We reasoned about properties involving noisy acting and sensing using Bel.

Syntactically, then, we will have the very same language as $\mathcal{ES}$, but with distinguished binary predicate fluents alt and l, together with an epistemic modality $B(\alpha : r)$, using r for rationals.[3] The latter formula is to be read as "α is believed with a probability r." In line with $\mathcal{ES}$, we can also introduce a companion only-knowing modality: $O(\alpha_1 : r_1, \ldots, \alpha_k : r_k)$, which is to be read as "all that is believed is: α_1 with probability $r_1, \ldots$, and α_k with probability r_k."

---

[3] As with $\mathcal{ES}$, we drop functional fluents for simplicity in presentation.

10.2.1 Semantics

To understand how the semantics for $\mathcal{DS}$ is to be designed from $\mathcal{ES}$ but whilst staying true to the situation calculus variant, consider that there are three key complications:

1. we need to be able to specify probabilities over *uncountably many possible* worlds in a well-defined manner;
2. we need to allow for qualitative uncertainty in an inherently quantitative account, beliefs *may not be characterizable* in terms of a single distribution; and
3. the effects of actions are nondeterministic, some more likely than others, and the changes to the state of affairs thereof are (possibly) not observable by the agent.

To address this, let us first define $\mathcal{Z}$, $\mathcal{P}$ and $\mathcal{W}$ as in $\mathcal{ES}$. In addition, we will require that at every world $w \in \mathcal{W}$,

- l behaves like a function, that is, for all a, z, there is exactly one rational $n \geq 0$ such that $w[l(a, n), z] = 1$ and for all rationals $n' \neq n$, $w[l(a, n'), z] = 0$; and
- alt is an equivalence relation (reflexive, symmetric, and transitive) for all z (for simplicity).

Then by a *distribution d* we mean a mapping from $\mathcal{W}$ to $\mathbb{R}^{\geq 0}$ (the set of non-negative reals) and an *epistemic state e* is now any set of distributions. So, basically every world is considered accessible in principle, but worlds inconsistent with what the agent is declared to know initially will clearly get a d-value of 0. By allowing e to be a set of such functions, we are enabling multiple distributions. Note, of course, d as defined above does not quite guarantee it is a measurable function, which will be addressed by means of a normalization condition below.

In fact, to prepare for the semantics, we will need four notational devices. First, we extend the application of l to sequences. We define $l^* : \mathcal{W} \times \mathcal{Z} \mapsto \mathbb{R}^{\geq 0}$ as follows:

- $l^*(w, \langle\rangle) = 1$ for every $w \in \mathcal{W}$;
- $l^*(w, z \cdot r) = l^*(w, z) \times n$ where $w[l(r, n), z] = 1$.

Second, after intending to execute a sequence of actions, the agent needs to also consider those sequences that are possibly the actual outcomes. For this, we define *action sequence observational indistinguishability* as follows. Given any world w, we define $z \sim_w z'$:

- $\langle\rangle \sim_w z'$ iff $z' = \langle\rangle$;
- $z \cdot r \sim_w z'$ iff $z' = z^* \cdot r^*$, $z \sim_w z^*$ and $w[alt(r, r^*), z] = 1$.

Since alt is an equivalence relation, it can be shown that $\sim_w$ is an equivalence relation.

Third, after actions, we will restrict ourselves to *compatible* worlds that agree on observational indistinguishability. We write $w \approx_{alt} w'$ iff:

- for all a, a', z, $w[oi(a, a'), z] = w'[alt(a, a'), z]$.

Fourth, to extend the applicability of *Poss* for action sequences, we proceed as follows. Define $Exec(z)$ for any $z \in \mathcal{Z}$ inductively:

- for $z = \langle \rangle$, $Exec(z)$ denotes *true*;
- for $z = a \cdot z'$, $Exec(z)$ denotes $Poss(a) \wedge [a]Exec(z')$.

We are finally ready for the semantics. To obtain a well-defined sum over uncountably many worlds, we will introduce some conditions for distributions used for evaluating epistemic operators. We define NORM, EQUAL, BOUND for any d and any set $\mathcal{V} = \{(w_1, z_1), (w_2, z_2), \ldots\}$ as follows:

1. for any $\mathcal{U} \subseteq \mathcal{V}$, NORM$(d, \mathcal{U}, \mathcal{V}, r)$ iff $\exists b \neq 0$ such that EQUAL$(d, \mathcal{U}, b \times r)$ and EQUAL$(d, \mathcal{V}, b)$.
2. EQUAL$(d, \mathcal{V}, r)$ iff BOUND$(d, \mathcal{V}, r)$ and there is no $r' < r$ such that BOUND$(d, \mathcal{V}, r')$ holds.
3. BOUND$(d, \mathcal{V}, r)$ iff $\neg \exists k, (w_1, z_1), \ldots, (w_k, z_k) \in \mathcal{V}$ such that

$$\sum_{i=1}^{k} d(w_i) \times l^*(w_i, z_i) > r.$$

Intuitively, given NORM$(d, \mathcal{U}, \mathcal{V}, r)$, r can be seen as the *normalization* of the weights of worlds in $\mathcal{U}$ in relation to the set of worlds $\mathcal{V}$ as accorded by d. Here, EQUAL$(d, \mathcal{V}, b)$ expresses that the weight accorded to the worlds in $\mathcal{V}$ is b, and finally BOUND$(d, \mathcal{V}, b)$ ensures the weight of worlds in $\mathcal{V}$ is bounded by b. In essence, although $\mathcal{W}$ is uncountable, the conditions BOUND and EQUAL admit a well-defined summation of the weights on worlds.

Truth in $\mathcal{DS}$ is defined w.r.t. triples (e, w, z), where the definition for atoms and connectives are as in $\mathcal{ES}$; for example:

- $e, w, z \models p$ iff p is an atom and $w[p, z] = 1$;
- $e, w, z \models \alpha \wedge \beta$ iff $e, w, z \models \alpha$ and $e, w, z \models \beta$.
- $e, w, z \models [a]\alpha$ iff $e, w, z \cdot a \models \alpha$.

For interpreting the epistemic modality, given a triple (e, w, z), let $\mathcal{W}_\alpha = \{(w', z') \mid z' \sim_w z, w' \approx_{alt} w$, and $e, w', \langle \rangle \models [z']\alpha \wedge Exec(z')\}$. That is, these are the pairs of worlds and executable action sequences that agree on *alt* with w, and where α holds. Then:

- $e, w, z \models B(\alpha : r)$ iff $\forall d \in e$, NORM$(d, \mathcal{W}_\alpha, \mathcal{W}_{true}, r)$.
- $e, w, z \models O(\alpha_1 : r_1, \ldots, \alpha_k : r_k)$ iff for all d, $d \in e$ iff NORM$(d, \mathcal{W}_{\alpha_1}, \mathcal{W}_{true}, r_1), \ldots$
  and NORM$(d, \mathcal{W}_{\alpha_k}, \mathcal{W}_{true}, r_k)$.

We define $\mathcal{D} \models \alpha$ as we would for $\mathcal{ES}$.

Putting it all together, the $\mathcal{DS}$ framework can be summarized as follows:

- worlds are trees that interpret properties initially as well as after any sequence of actions;
- the agent's mental state is defined in terms of a set of distributions over worlds;
- the agent's belief in a formula α is a number obtained by summing the d-value of worlds where α is true; and, every d in the agent's mental state should agree on that number.

Readers may notice that this is very much in the spirit of the situation calculus exposition. We will now consider some interesting properties of the logic which admit fairly straightforward arguments.

10.2.2 Properties

What kind of properties are reasonable for a probabilistic $\mathcal{ES}$? In the very least, we should expect the properties of knowledge, suitably adjusted, to also hold. But what properties can we show about how B behaves? Moreover, what about the relation between $\mathcal{DS}$ and $\mathcal{ES}$?

Let us proceed in reverse order. $\mathcal{ES}$ worlds are precisely $\mathcal{DS}$ worlds, so let us suppress the differences in distinguished symbols. Let $\Box \forall a, a'$ $(alt(a, a') \equiv a = a')$ and $\Box \forall a, u$ $(l(a, u) \equiv u = 1)$ be lumped together as Ω. These say that a is only alt-related to itself, and that the likelihood of every action is 1. Conversely, let $\Gamma = \{\Box \forall a \ (SF(a) \equiv true)\}$, where actions have trivial sensing results. Let us also use $K\alpha$ in $\mathcal{DS}$ to mean $B(\alpha : 1)$, and likewise use $O\alpha$ to mean $O(\alpha : 1)$. Then:

Theorem 10.4 *Let α be any $\mathcal{ES}$-formula. Then $\Omega \supset \alpha$ is valid in $\mathcal{DS}$ iff $\Gamma \supset \alpha$ is valid in $\mathcal{ES}$.*

So, basically, $\mathcal{ES}$ is part of $\mathcal{DS}$.[4]

Now let us consider properties for B. For starters, B is well-behaved in terms of semantical equivalence; for any α, β such that $\alpha \equiv \beta$, we have $B(\alpha : r) \equiv B(\beta : r)$. But beyond that, things like the addition law of probability are easily shown to be valid:

---

[4] Recall that in the probabilistic situation calculus there is no SF function. That is, on obtaining a sensing outcome the worlds that disagree with the outcome are not discarded but merely have their weights adjusted. Nonetheless, we suspect this result can be further extended to also handle non-trivial sensing outcomes using extra axioms to align l and SF.

$$\Box(B(\alpha:r) \land B(\beta:r') \land B(\alpha \land \beta:r'') \supset B(\alpha \lor \beta:r+r'-r'')).$$

That is, for all action sequences (exactly like how we positioned properties in $\mathcal{ES}$), the probability of the union of two events is obtained from the probabilities of the individual events while discounting for overlap. The proof for such properties is not dissimilar to classical probability theory, except that we need to argue in terms of the sets $\mathcal{W}_\alpha$, $\mathcal{W}_\beta$, $\mathcal{W}_{\alpha \land \beta}$ and $\mathcal{W}_{\alpha \lor \beta}$.

Finally, from the relationship between $\mathcal{ES}$ and $\mathcal{DS}$, it is perhaps clear that every K-related property in $\mathcal{ES}$ has a natural analogue in $\mathcal{DS}$. Consider positive introspection. In $\mathcal{DS}$, we have:

$$\models \Box(K\alpha \supset KK\alpha),$$

as one would in $\mathcal{ES}$, but this is true also with degrees of belief:

$$\models \Box(B(\alpha:r) \supset KB(\alpha:r)).$$

That is, the agent knows the degree to which it believes a formula α.

There are, of course, many other properties that can be explored, such as the relation between O, K and B, but we hope the readers get the general idea that the framework is very amenable to such types of inquiry.

10.2.3 Axiomatization: The Robot with Noisy Effectors

Basic action theories in $\mathcal{DS}$ will include the usual ingredients on the executability of actions and their effects as in $\mathcal{ES}$, but will additionally include axioms about observational indistinguishability and action likelihoods.

Let us revisit the robot example, with some minor simplifications. The robot is trying to move towards the wall, and as usual, we use $move(x, y)$ to model a nondeterministic outcome y when the robot intends to move by x.

In terms of preconditions, the amount the robot can movement is limited by the distance to the wall, but suppose the sonar is always executable:

$$\Box Poss(a) \equiv \exists x, y, u(a = move(x, y) \land h(u) \land y \le u) \lor \exists z(a = sonar(z) \land true).$$

As before, the distance to the wall, given by fluent h, is reduced by the actual amount moved:

$$\Box[a]h(u) \equiv \exists x, y(a = move(x, y) \land h(u + y)) \lor \forall x, y (a \ne move(x, y) \land h(u)).$$

For capturing nondeterminism, let us suppose the move action is noisy only in the sense of the intended argument not always succeeding:

$$\Box alt(a, a') \equiv \exists x, y, z(a = move(x, y) \land a' = move(x, z)) \lor (a = sonar(z) \land a' = a).$$

The likelihood for the move and the sensing action is given by:

$$\Box l(a, u) \equiv \exists x, y, z (a = sonar(z) \wedge u = \Theta(h, z, .8, .1)) \vee$$
$$(a = move(x, y) \wedge u = \Theta(x, y, .6, .2)) \vee$$
$$(a \neq sonar(z) \wedge a \neq move(x, y) \wedge u = 0).$$

We use Θ for the likelihood accorded to differences in the intended argument and the actual value:

$$\Theta(u, v, c, d) = \begin{cases} c & \text{if } u = v \\ d & \text{if } |u - v| = 1 \\ 0 & \text{otherwise} \end{cases}$$

We lump these axioms together as Σ. Finally, for the robot's initial knowledge, assume a standard discrete uniform distribution on h on the range $\{2, 3, 4\}$. Suppose this and the basic action theory is all that is known; we then have:

$$O(h(2): 1/3; h(3): 1/3; h(4): 1/3; \Sigma: 1).$$

Example 10.5 The following are the entailments of the above sentence:

1. $B(h(5): 0)$
   By means of defining a probability distribution over 3 possible values for h, other values are impossible.
2. $[sonar(2)]B(h(4): 0) \wedge [sonar(2)]B(h(2): 8/9)$
   Obtaining a reading of 2 on the sensor means that being 4 units away is no longer possible, whereas the agent's confidence in being 2 units away from the wall increases.
3. $[sonar(2)][move(1, 1)]B(h(2): 8/45)$
   Consider that, in case of an exact move, the probability of being 2 units away would have been 0. See Fig. 10.1 for the degrees of belief in other values of h.

Example 10.6 Let us now consider a slightly more interesting variant. Assume the same dynamic axioms as above (given by Σ), but imagine the agent to only have qualitative uncertainty about h's value:

Fig. 10.1 Distribution on h values from the example

h values	2	3	4	5
initially	1/3	1/3	1/3	0
after *sonar(2)*	8/9	1/9	0	0
after *move(1,1)*	8/45	5/9	11/45	1/45

$$O(\exists u\ (h(u) \wedge u > 1) \wedge \Sigma : 1).$$

So, the agent considers infinitely many h values possible. The following are its entailments:

1. $B(h(1): 0) \wedge \neg B(h(4): 0)$
   Values for $h > 1$ cannot be ruled out. Here, only knowing allows us to infer non-beliefs.
2. $[sonar(2)]B(h(4): 0)$
   The nature of the likelihood axioms for the sensor is such that obtaining a reading of 2 eliminates infinitely many possibilities.

It is worth noting that the following sentence, for example, is not entailed:

3. $[sonar(2)]\exists u, v\ B(h(2): u) \wedge B(h(3): v) \wedge u > v$
   A sensor reading of 2 means that the robot is either 2 or 3 units away from the wall, as the prior on being 1 unit away is 0. Despite the reading favoring the case for the robot being two units away from the wall, qualitative uncertainty about h's value means there are distributions where $h(2)$ has a low or even 0 prior probability, and therefore, it does not follow that the degree of belief in $h(2)$ necessarily trumps that in $h(3)$.

10.3   Notes

For a concise introduction to modal logic, see [44]. For studies on modal logic in philosophy and formal verification, see [50, 56, 61, 73].

Before discussing related work, let us recap that we have presented a modal language corresponding to the situation calculus with degrees of belief from the previous chapters. The arguments from the start of the chapter for the need for a modal reconstruction of the situation calculus are taken from [136]. It introduces the logic $\mathcal{ES}$ as a modal variant of the epistemic situation calculus, with both the knowing as well as the only-knowing modal operator. The semantics for the only knowing operator is based on [148]. Translating between $\mathcal{ES}$ formulas and classical epistemic situation calculus formulas was done in [137]. Finally, for discussions on measurable functions, see [31, 99].[5]

Clearly, many other results need yet to be recast for $\mathcal{DS}$, including the correctness of the translation as established for $\mathcal{ES}$ in [137], support for continuous distributions, regression, progression, and the semantics of programs. We do not expect this recasting to be very problematic, but rather for it to provide further clarity on beliefs and meta-beliefs in a clean

---

[5] It is worth noting that when defining the modal operator for degrees of belief, the conditions are set up so as to lead to epistemic states that are measurable functions [20]. Interestingly, like with *Bel* in the non-modal situation calculus, we only define the summation operator, and if the sum is defined, we are able to reason about beliefs. So our conditions, in a sense, recreate that simplicity of the BHL proposal in a modal language.

possible-worlds framework. In fact, there is already work on some of these directions [159, 220], and as expected the modal variant leads to a crisper study of the changing degrees of beliefs.

As mentioned in previous chapters, reasoning about probabilities is widely studied in the logical literature [82, 101, 174]. Although we have already contrasted our framework to these, a further distinguishing feature of $\mathcal{DS}$ is the modality for only knowing.[6]

Reasoning about knowledge and probability has appeared in a number of works, of course, in computer science [72], game theory [109, 170], knowledge representation [224], and program analysis [106, 129]. Properties discussed in this chapter, such as introspection and additivity, are also well studied [4]. Notably, the work of Fagin and Halpern [72] is very similar to the thrust of the probabilistic situation calculus and $\mathcal{DS}$, although it is propositional and does not consider actions.

In previous chapters, we also articulated the differences between contemporary, less expressive but tractable probabilistic logical models from the machine learning literature and the probabilistic situation calculus, in that the latter supports full first-order logic and is embedded in a general theory of action. The logic $\mathcal{DS}$ clearly inherits the latter's features, but additionally allows for an easier modelling and reasoning of meta-beliefs.

There are at least two interesting avenues for future work. First, only-knowing has been extended to multiple agents in dynamic domains previously [18], and adding probabilities to such languages would yield a multi-agent $\mathcal{DS}$. Such a language would allow us to further capture and unify logic and probability in multi-agent systems. Second, by considering a ALLEGRO-type language in $\mathcal{DS}$, we would be laying the groundwork for considering the verification of temporal properties, such as safety and liveness. We might be able to draw ideas from previous investigations of this sort for GOLOG and knowledge-based GOLOG [47, 49, 50, 234].

---

[6] Only knowing is related to notions such as *minimal knowledge* [103] and *total knowledge* [185]. See [152] for discussions. In earlier work, Gabaldon and Lakemeyer [81] considered a logic of only knowing and probability by meta-linguistically enforcing finitely many equivalence classes for possible worlds. Consequently, quantification also ranges over a finite set. In contrast, by using the constraints of NORM, for example, we could allow distributions on uncountably many worlds. In a game theory context, Halpern and Pass [105] have considered a (propositional) version of only knowing with probability distributions.

Conclusions

11

Context is to data what water is to a dolphin.
—**Dan Simmons**, *Olympos*

11.1 Summary

Starting with the motto of "representation first, acquisition second", we have worked through a formal language for unifying logic, probability and actions. As it is a first-order representation, the language is powerful and very expressive. The situation calculus was already shown to capture intricate notions of time, belief update, belief revision, decision theory, concurrency, and so on. By now showing a simple integration with probabilities, we now have a general language for degrees of belief and noisy sensing and acting.

The main advantage of our logical account is that it allows a specification of belief that can be partial or incomplete, in keeping with whatever information is available about the application domain. It does not require specifying a prior distribution over some random variables from which posterior distributions are then calculated. Nor does it require specifying the conditional independences among random variables and how these dependencies change as the result of actions, as in the temporal extensions to Bayesian networks. In our model, some logical constraints are imposed on the initial state of belief. These constraints may be compatible with one or very many initial distributions and sets of independence assumptions. All the properties of belief will then follow at a corresponding level of specificity.

As suggested in the first chapter, readers not wishing to work with the situation calculus would still benefit from the inquiries in this book. In the very least, it is a general-purpose first-order logical language for probabilistic knowledge and acting, accompanied by results on projection and programming. A modal reconstruction of the situation calculus yielded

© The Author(s), under exclusive license to Springer Nature Switzerland AG 2023
V. Belle, *Toward Robots That Reason: Logic, Probability & Causal Laws*,
Synthesis Lectures on Artificial Intelligence and Machine Learning,
https://doi.org/10.1007/978-3-031-21003-7_11

a language that is perhaps easier to follow for those familiar with dynamic and program logics, and could be a starting point for the reader to define their own formal apparatus for logic + probability + actions.

11.2   What About Automated Planning?

In our book, we have been suspiciously silent about planning and/or program synthesis, that is, the provision of an algorithm that produces a sequence of actions that achieves the goal. Automated planning gets significantly more interesting with beliefs, where actions have to be interleaved with sensing to know more about the world before the goal can be achieved. Add probabilities to the mix, and we now have to also account for actions failing and unreliable observations. Think, for example, of providing an algorithm that outputs a program like the one we discussed with ALLEGRO.

Fortunately, we do have something to say about the matter beyond idle speculation. Note that for the two decades, there has considerable work on synthesizing plans with noisy acting, as seen for partially observable Markov decision processes (POMDPs). While that requires a reward function, more recently, there is also work on synthesizing noisy plans for belief-level goals (i.e., epistemic formulas) using a regression operator. What about non-sequential plans, such as programs, involving recursion? Here too, there are approaches that can deal with noise. However, there is quite a bit of nuance in relating these results to what is possible with the full language of the situation calculus considered in this book, and so we have the left the matter for a future edition where a comprehensive and unified treatment might be possible.

11.3   Outlook

The unification of logic + probability + actions is only one of the steps needed for rational commonsense agents. On a pragmatic level, for example, the entire book focused on a simple example—although this was justified in the need to better understand the nuances of the unification—further work is needed to see how a large-scale probabilistic knowledge base might work with noisy actions and observations. To ensure real-time behavior, an investigation on useful fragments is needed. And, as suggested in the first chapter, we need to allow the agents to acquire associative and causal knowledge from data. The data gathering process itself might involve, say, communication with multiple agents. Low-level data needs to translate to and be abstracted as high-level concepts involving time, space and causality. Such problems are very much at the heart of general-purpose, open-ended AI, that can reason with both data and context about the natural world.

If we are to draw inspiration from human cognition, Daniel Kahneman's so-called *System 1* versus *System 2* processing is a worthwhile model. This might be understood as the search for a framework that links experiential and reactive processing (learned behavior) with cogitative processing (reasoning, deliberation and introspection) in service of sophisticated machine intelligence. In that regard, there is clearly a need for integrating high-level conceptual knowledge with probabilistic low-level observations. The unification of logic, probability and actions could fill that need.

11.4 Notes

For planning solutions based on POMDP, see [118] as a starting point, and [85] for a broader discussion. Regression-based planning with noisy sensing is considered in [120, 183]. Recursive solutions for automated planning under noise is analyzed in [16, 40–42, 223]. For perspectives on connecting such recursive solutions to the probabilistic situation calculus, see [16].

Daniel Kahneman's distinction on the types of cognitive processing is summarized in [121]. Discussions on the desiderata for rational commonsense agents appears in books such as [150, 163, 180], among others.

Bibliography

1. C. E. Alchourròn, P. Gärdenfors, and D. Makinson. On the logic of theory change: Partial meet contraction and revision functions. *Journal of Symbolic Logic*, 50:510–530, 1985.
2. C. Anderson, P. Domingos, and D. Weld. Relational markov models and their application to adaptive web navigation. In *Proc. SIGKDD*, pages 143–152. ACM, 2002.
3. D. Appelt and K. Konolige. A practical nonmonotonic theory for reasoning about speech acts. In *Proc. ACL*, pages 170–178, 1988.
4. R. J. Aumann. Interactive epistemology II: probability. *Int. J. Game Theory*, 28(3):301–314, 1999.
5. F. Bacchus. *Representing and Reasoning with Probabilistic Knowledge*. MIT Press, 1990.
6. F. Bacchus, S. Dalmao, and T. Pitassi. Solving #SAT and Bayesian inference with backtracking search. *J. Artif. Intell. Res. (JAIR)*, 34:391–442, 2009.
7. F. Bacchus, A. J. Grove, J. Y. Halpern, and D. Koller. From statistical knowledge bases to degrees of belief. *Artificial intelligence*, 87(1-2):75–143, 1996.
8. F. Bacchus, J. Y. Halpern, and H. J. Levesque. Reasoning about noisy sensors and effectors in the situation calculus. *Artificial Intelligence*, 111(1–2):171 – 208, 1999.
9. J. A. Baier, C. Fritz, and S. A. McIlraith. Exploiting procedural domain control knowledge in state-of-the-art planners. In *Proc. ICAPS*, pages 26–33, 2007.
10. B. Banihashemi, G. De Giacomo, and Y. Lespérance. Abstraction in situation calculus action theories. In *Thirty-First AAAI Conference on Artificial Intelligence*, 2017.
11. C. Baral, T. Bolander, H. van Ditmarsch, and S. McIlrath. Epistemic Planning (Dagstuhl Seminar 17231). *Dagstuhl Reports*, 7(6):1–47, 2017.
12. C. Baral, M. Gelfond, and J. N. Rushton. Probabilistic reasoning with answer sets. *TPLP*, 9(1):57–144, 2009.
13. V. Batusov and M. Soutchanski. Situation calculus semantics for actual causality. In *Proceedings of the AAAI Conference on Artificial Intelligence*, volume 32, 2018.
14. V. Belle. On plans with loops and noise. In *Proceedings of the 17th International Conference on Autonomous Agents and MultiAgent Systems*, pages 1310–1317, 2018.
15. V. Belle. Logic meets learning: From aristotle to neural networks. In *Neuro-Symbolic Artificial Intelligence: The State of the Art*, pages 78–102. IOS Press, 2021.

© The Editor(s) (if applicable) and The Author(s), under exclusive license to Springer
Nature Switzerland AG 2023
V. Belle, *Toward Robots That Reason: Logic, Probability & Causal Laws*,
Synthesis Lectures on Artificial Intelligence and Machine Learning,
https://doi.org/10.1007/978-3-031-21005-7

16. V. Belle. Analyzing generalized planning under nondeterminism. *Artificial Intelligence*, page 103696, 2022.

17. V. Belle and B. Juba. Implicitly learning to reason in first-order logic. *Advances in Neural Information Processing Systems*, 32, 2019.

18. V. Belle and G. Lakemeyer. Multiagent only knowing in dynamic systems. *Journal of Artificial Intelligence Research*, 49, 2014.

19. V. Belle and G. Lakemeyer. On the progression of knowledge in multiagent systems. In *KR*, 2014.

20. V. Belle, G. Lakemeyer, and H. J. Levesque. A first-order logic of probability and only knowing in unbounded domains. In *Proc. AAAI*, 2016.

21. V. Belle and H. Levesque. Foundations for generalized planning in unbounded stochastic domains. In *KR*, 2016.

22. V. Belle and H. J. Levesque. Reasoning about continuous uncertainty in the situation calculus. In *Proc. IJCAI*, 2013.

23. V. Belle and H. J. Levesque. Reasoning about probabilities in dynamic systems using goal regression. In *Proc. UAI*, 2013.

24. V. Belle and H. J. Levesque. How to progress beliefs in continuous domains. In *Proc. KR*, 2014.

25. V. Belle and H. J. Levesque. A logical theory of robot localization. In *Proc. AAMAS*, 2014.

26. V. Belle and H. J. Levesque. PREGO: An Action Language for Belief-Based Cognitive Robotics in Continuous Domains. In *Proc. AAAI*, 2014.

27. V. Belle and H. J. Levesque. Allegro: Belief-based programming in stochastic dynamical domains. In *IJCAI*, 2015.

28. V. Belle and H. J. Levesque. Reasoning about discrete and continuous noisy sensors and effectors in dynamical systems. *Artificial Intelligence*, 262:189–221, 2018.

29. V. Belle, A. Passerini, and G. Van den Broeck. Probabilistic inference in hybrid domains by weighted model integration. In *IJCAI*, 2015.

30. D. Bertsekas and J. Tsitsiklis. *Introduction to probability*. Athena Scientific, 2nd edition edition, 2008.

31. P. Billingsley. *Probability and Measure*. Wiley-Interscience, 3 edition, Apr. 1995.

32. C. Boutilier, T. Dean, and S. Hanks. Decision-theoretic planning: Structural assumptions and computational leverage. *Journal of Artificial Intelligence Research*, 11(1):94, 1999.

33. C. Boutilier, R. Reiter, and B. Price. Symbolic dynamic programming for first-order MDPs. In *Proc. IJCAI*, pages 690–697, 2001.

34. C. Boutilier, R. Reiter, M. Soutchanski, and S. Thrun. Decision-theoretic, high-level agent programming in the situation calculus. In *Proc. AAAI*, pages 355–362, 2000.

35. G. E. P. Box and G. C. Tiao. *Bayesian inference in statistical analysis*. Addison-Wesley, 1973.

36. X. Boyen and D. Koller. Tractable inference for complex stochastic processes. In *Proc. UAI*, pages 33–42, 1998.

37. R. Brachman and H. Levesque. *Knowledge representation and reasoning*. Morgan Kaufmann Pub, 2004.

38. W. Burgard, A. B. Cremers, D. Fox, D. Hähnel, G. Lakemeyer, D. Schulz, W. Steiner, and S. Thrun. Experiences with an interactive museum tour-guide robot. *Artif. Intell.*, 114(1-2):3–55, 1999.

39. R. Carnap. *Logical foundations of probability*. Routledge and Kegan Paul London, 1951.

40. K. Chatterjee and M. Chmelík. Pomdps under probabilistic semantics. *Artificial Intelligence*, 221:46–72, 2015.

41. K. Chatterjee, M. Chmelik, and J. Davies. A symbolic sat-based algorithm for almost-sure reachability with small strategies in pomdps. In *Proceedings of the AAAI Conference on Artificial Intelligence*, volume 30, 2016.

42. K. Chatterjee, M. Chmelik, R. Gupta, and A. Kanodia. Qualitative analysis of pomdps with temporal logic specifications for robotics applications. In *2015 IEEE International Conference on Robotics and Automation (ICRA)*, pages 325–330. IEEE, 2015.

43. M. Chavira and A. Darwiche. On probabilistic inference by weighted model counting. *Artif. Intell.*, 172(6-7):772–799, 2008.

44. B. Chellas. *Modal logic*. Cambridge University Press, 1980.

45. J. Choi, A. Guzman-Rivera, and E. Amir. Lifted relational kalman filtering. In *Proc. IJCAI*, pages 2092–2099, 2011.

46. A. Cimatti, M. Pistore, M. Roveri, and P. Traverso. Weak, strong, and strong cyclic planning via symbolic model checking. *Artificial Intelligence*, 147(1–2):35 – 84, 2003.

47. J. Claßen. Symbolic verification of Golog programs with first-order BDDs. In M. Thielscher, F. Toni, and F. Wolter, editors, *Proceedings of the Sixteenth International Conference on Principles of Knowledge Representation and Reasoning (KR 2018)*, pages 524–528. AAAI Press, 2018.

48. J. Claßen and G. Lakemeyer. Foundations for knowledge-based programs using ES. In *Proc. KR*, pages 318–328, 2006.

49. J. Claßen and G. Lakemeyer. A logic for non-terminating golog programs. In *KR*, pages 589–599, 2008.

50. J. Claßen, M. Liebenberg, G. Lakemeyer, and B. Zarrieß. Exploring the boundaries of decidable verification of non-terminating Golog programs. In *AAAI*, 2014.

51. I. J. Cox. Blanche-an experiment in guidance and navigation of an autonomous robot vehicle. *Robotics and Automation, IEEE Transactions on*, 7(2):193–204, 1991.

52. F. G. Cozman. Credal networks. *Artificial Intelligence*, 120(2):199 – 233, 2000.

53. G. Da Prato. *An Introduction to Infinite-Dimensional Analysis*. Universitext. Springer, 2006.

54. A. Darwiche and M. Goldszmidt. Action networks: A framework for reasoning about actions and change under uncertainty. In *Proc. UAI*, pages 136–144, 1994.

55. E. Davis. *Representations of commonsense knowledge*. Morgan Kaufmann, 2014.

56. L. de Alfaro. Quantitative verification and control via the mu-calculus. In *CONCUR*, pages 102–126, 2003.

57. G. De Giacomo and H. Levesque. Two approaches to efficient open-world reasoning. In *Logic-based artificial intelligence*, pages 59–78. Kluwer Academic Publishers, Norwell, MA, USA, 2000.

58. L. De Raedt and K. Kersting. Statistical relational learning. In *Encyclopedia of Machine Learning*, pages 916–924. Springer, 2011.

59. T. Dean and K. Kanazawa. Probabilistic temporal reasoning. In *Proc. AAAI*, pages 524–529, 1988.

60. T. Dean and M. Wellman. *Planning and control*. Morgan Kaufmann Publishers Inc., 1991.

61. J. A. DeCastro and H. Kress-Gazit. Synthesis of nonlinear continuous controllers for verifiably correct high-level, reactive behaviors. *I. J. Robotic Res.*, 34(3):378–394, 2015.

62. J. P. Delgrande and H. J. Levesque. Belief revision with sensing and fallible actions. In *Proc. KR*, 2012.

63. F. Dellaert, D. Fox, W. Burgard, and S. Thrun. Monte carlo localization for mobile robots. In *Robotics and Automation, 1999. Proceedings. 1999 IEEE International Conference on*, volume 2, pages 1322–1328. IEEE, 1999.

64. R. Demolombe. Belief change: from situation calculus to modal logic. In *Proc. Nonmonotonic Reasoning, Action, and Change (NRAC)*, 2003.

65. R. Demolombe, A. Herzig, and I. Varzinczak. Regression in modal logic. *Journal of Applied Non-Classical Logics*, 13(2):165–185, 2003.

66. M. Denecker, M. Bruynooghe, and V. Marek. Logic programming revisited: Logic programs as inductive definitions. *ACM Transactions on Computational Logic*, 2(4):623–654, 2001.

67. M. Denecker and E. Ternovska. Inductive situation calculus. *Artificial Intelligence*, 171(5–6):332 – 360, 2007.

68. P. Domingos and W. A. Webb. A tractable first-order probabilistic logic. In *AAAI*, 2012.

69. H. Enderton. *A mathematical introduction to logic*. Academic press New York, 1972.

70. L. Eronen and H. Toivonen. Biomine: predicting links between biological entities using network models of heterogeneous databases. *BMC bioinformatics*, 13(1):1–21, 2012.

71. C. J. Ewin, A. R. Pearce, and S. Vassos. Transforming situation calculus action theories for optimised reasoning. In *Principles of Knowledge Representation and Reasoning: Proceedings of the Fourteenth International Conference, KR 2014, Vienna, Austria, July 20-24, 2014*, 2014.

72. R. Fagin and J. Y. Halpern. Reasoning about knowledge and probability. *J. ACM*, 41(2):340–367, 1994.

73. R. Fagin, J. Y. Halpern, Y. Moses, and M. Y. Vardi. *Reasoning About Knowledge*. MIT Press, 1995.

74. Y. Fan, M. Cai, N. Li, and Y. Liu. A first-order interpreter for knowledge-based golog with sensing based on exact progression and limited reasoning. In *Proc. AAAI*, 2012.

75. A. Ferrein and G. Lakemeyer. Logic-based robot control in highly dynamic domains. *Robotics and Autonomous Systems*, 56(11):980–991, 2008.

76. D. Fierens, G. V. den Broeck, I. Thon, B. Gutmann, and L. D. Raedt. Inference in probabilistic logic programs using weighted CNF's. In *UAI*, pages 211–220, 2011.

77. A. Finzi and T. Lukasiewicz. Structure-based causes and explanations in the independent choice logic. In *Proceedings of the Nineteenth conference on Uncertainty in Artificial Intelligence*, pages 225–323, 2002.

78. D. Fox, J. Hightower, L. Liao, D. Schulz, and G. Borriello. Bayesian filtering for location estimation. *Pervasive Computing, IEEE*, 2(3):24–33, 2003.

79. M. Fox and D. Long. Modelling mixed discrete-continuous domains for planning. *J. Artif. Intell. Res. (JAIR)*, 27:235–297, 2006.

80. C. Fritz and S. A. McIlraith. Computing robust plans in continuous domains. In *Proc. ICAPS*, pages 346–349, 2009.

81. A. Gabaldon and G. Lakemeyer. ESP: A logic of only-knowing, noisy sensing and acting. In *Proc. AAAI*, pages 974–979, 2007.

82. H. Gaifman. Concerning measures in first order calculi. *Israel Journal of Mathematics*, 2(1):1–18, 1964.

83. A. S. d. Garcez, K. Broda, D. M. Gabbay, et al. *Neural-symbolic learning systems: foundations and applications*. Springer Science & Business Media, 2002.

84. X. Ge and J. Renz. Representation and reasoning about general solid rectangles. In *IJCAI*, 2013.

85. H. Geffner and B. Bonet. *A Concise Introduction to Models and Methods for Automated Planning*. Morgan and Claypool Publishers, 2013.

86. M. Gelfond and V. Lifschitz. Representing action and change by logic programs. *The Journal of Logic Programming*, 17(2-4):301–321, 1993.

87. M. Gelfond and V. Lifschitz. Special issue: Non-monotonic reasoning and logic programming representing action and change by logic programs. *The Journal of Logic Programming*, 17(2):301 – 321, 1993.

88. J. Gerbrandy and W. Groeneveld. Reasoning about information change. *J. of Logic, Lang. and Inf.*, 6(2):147–169, Apr. 1997.

89. L. Getoor, N. Friedman, D. Koller, and B. Taskar. Learning probabilistic models of relational structure. In *ICML*, pages 170–177, 2001.

90. L. Getoor and B. Taskar, editors. *An Introduction to Statistical Relational Learning*. MIT Press, 2007.

91. L. Getoor and B. Taskar. Introduction to statistical relational learning (adaptive computation and machine learning). 2007.

92. M. Ghallab, D. Nau, and P. Traverso. *Automated Planning: theory and practice*. Morgan Kaufmann Publishers, 2004.

93. A. M. Glenberg and D. A. Robertson. Symbol grounding and meaning: A comparison of high-dimensional and embodied theories of meaning. *Journal of memory and language*, 43(3):379–401, 2000.

94. V. Gogate and P. Domingos. Formula-based probabilistic inference. In *Proc. UAI*, pages 210–219, 2010.

95. N. D. Goodman, V. K. Mansinghka, D. M. Roy, K. Bonawitz, and J. B. Tenenbaum. Church: a language for generative models. In *Proc. UAI*, pages 220–229, 2008.

96. H. Grosskreutz and G. Lakemeyer. ccgolog – a logical language dealing with continuous change. *Logic Journal of the IGPL*, 11(2):179–221, 2003.

97. H. Grosskreutz and G. Lakemeyer. Probabilistic complex actions in golog. *Fundam. Inform.*, 57(2-4):167–192, 2003.

98. H. Hajishirzi and E. Amir. Reasoning about deterministic actions with probabilistic prior and application to stochastic filtering. In *Proc. KR*, 2010.

99. P. Halmos. Measure theory. *Van Nostrad Reinhold Company*, 1950.

100. J. Halpern. An analysis of first-order logics of probability. *Artificial Intelligence*, 46(3):311–350, 1990.

101. J. Y. Halpern. *Reasoning about Uncertainty*. MIT Press, 2003.

102. J. Y. Halpern. *Actual causality*. MiT Press, 2016.

103. J. Y. Halpern and Y. Moses. Towards a theory of knowledge and ignorance: Preliminary report. In *Proc. NMR*, pages 125–143, 1984.

104. J. Y. Halpern and Y. Moses. A procedural characterization of solution concepts in games. *JAIR*, 49:143–170, 2014.

105. J. Y. Halpern and R. Pass. A logical characterization of iterated admissibility. In *Proc. TARK*, pages 146–155, 2009.

106. J. Y. Halpern and M. R. Tuttle. Knowledge, probability, and adversaries. *J. ACM*, 40:917–960, 1993.

107. J. Y. Halpern and M. Y. Vardi. Model checking vs. theorem proving: A manifesto. In J. F. Allen, R. Fikes, and E. Sandewall, editors, *KR*, pages 325–334. Morgan Kaufmann, 1991.

108. S. Harnad. The symbol grounding problem. *Physica D: Nonlinear Phenomena*, 42(1-3):335–346, 1990.

109. A. Heifetz and P. Mongin. Probability logic for type spaces. *Games and Economic Behavior*, 35(1-2):31 – 53, 2001.

110. C. S. Herrmann and M. Thielscher. Reasoning about continuous processes. In *AAAI/IAAI, Vol. 1*, pages 639–644, 1996.

111. A. Herzig, J. Lang, and P. Marquis. Action representation and partially observable planning using epistemic logic. In *Proc. IJCAI*, pages 1067–1072, 2003.

112. J. Hintikka. *Knowledge and belief: an introduction to the logic of the two notions*. Cornell University Press, 1962.

113. C. Hitchcock. Causality: Models, reasoning and inference, 2001.

114. M. Hopkins and J. Pearl. Clarifying the usage of structural models for commonsense causal reasoning. In *Proceedings of the AAAI Spring Symposium on Logical Formalizations of Commonsense Reasoning*, pages 83–89. AAAI Press Menlo Park, CA, 2003.

115. M. Hopkins and J. Pearl. Causality and counterfactuals in the situation calculus. *Journal of Logic and Computation*, 17(5):939–953, 2007.

116. Y. Hu and G. De Giacomo. A generic technique for synthesizing bounded finite-state controllers. In *ICAPS*, 2013.

117. Y. Hu and H. J. Levesque. A correctness result for reasoning about one-dimensional planning problems. In *IJCAI*, pages 2638–2643, 2011.

118. L. P. Kaelbling, M. L. Littman, and A. R. Cassandra. Planning and acting in partially observable stochastic domains. *Artificial Intelligence*, 101(1–2):99 – 134, 1998.

119. L. P. Kaelbling and T. Lozano-Pérez. Unifying perception, estimation and action for mobile manipulation via belief space planning. In *Proc. ICRA*, pages 2952–2959, 2012.

120. L. P. Kaelbling and T. Lozano-Pérez. Integrated task and motion planning in belief space. *I. J. Robotic Res.*, 32(9-10):1194–1227, 2013.

121. D. Kahneman. *Thinking, fast and slow*. Macmillan, 2011.

122. R. F. Kelly and A. R. Pearce. Complex epistemic modalities in the situation calculus. In *KR*, 2008.

123. S. M. Khan and Y. Lespérance. Knowing why—on the dynamics of knowledge about actual causes in the situation calculus. In *Proceedings of the 20th International Conference on Autonomous Agents and MultiAgent Systems*, pages 701–709, 2021.

124. D. Koller and N. Friedman. *Probabilistic Graphical Models - Principles and Techniques*. MIT Press, 2009.

125. D. Koller and A. Pfeffer. Probabilistic frame-based systems. In *AAAI/IAAI*, pages 580–587, 1998.

126. B. Kooi. Probabilistic dynamic epistemic logic. *Journal of Logic, Language and Information*, 12(4):381–408, 2003.

127. R. Kowalski and M. Sergot. A logic-based calculus of events. *New Generation Computing*, 4:67–95, 1986.

128. R. A. Kowalski. Predicate logic as programming language. In *IFIP Congress*, pages 569–574, 1974.

129. D. Kozen. Semantics of probabilistic programs. *Journal of Computer and System Sciences*, 22(3):328 – 350, 1981.

130. S. Kripke. A completeness theorem in modal logic. *Journal of Symbolic Logic*, 24(1):1–14, 1959.

131. S. Kripke. Semantical considerations on modal logic. *Acta Philosophica Fennica*, 16:83–94, 1963.

132. F. R. Kschischang, B. J. Frey, and H. Loeliger. Factor graphs and the sum-product algorithm. *IEEE Transactions on Information Theory*, 47(2):498–519, 2001.

133. N. Kushmerick, S. Hanks, and D. Weld. An algorithm for probabilistic planning. *Artificial Intelligence*, 76(1):239–286, 1995.

134. G. Lakemeyer. The situation calculus: A case for modal logic. *Journal of Logic, Language and Information*, 19(4):431–450, 2010.

135. G. Lakemeyer and H. J. Levesque. Evaluation-based reasoning with disjunctive information in first-order knowledge bases. In *Proc. KR*, pages 73–81, 2002.

136. G. Lakemeyer and H. J. Levesque. Situations, Si! situation terms, No! In *Proc. KR*, pages 516–526, 2004.

137. G. Lakemeyer and H. J. Levesque. Semantics for a useful fragment of the situation calculus. In *IJCAI*, pages 490–496, 2005.

138. G. Lakemeyer and H. J. Levesque. Cognitive robotics. In *Handbook of Knowledge Representation*, pages 869–886. Elsevier, 2007.

139. G. Lakemeyer and H. J. Levesque. A semantical account of progression in the presence of defaults. In *IJCAI*, pages 842–847, 2009.
140. J. Lang and B. Zanuttini. Knowledge-based programs as plans - the complexity of plan verification. In *Proc. ECAI*, pages 504–509, 2012.
141. T. Lang, M. Toussaint, and K. Kersting. Exploration in relational domains for model–based reinforcement learning. *Journal of Machine Learning Research (JMLR)*, 13(Dec):3691-3734, 2012.
142. N. Laverny and J. Lang. From knowledge-based programs to graded belief-based programs, part i: On-line reasoning. *Synthese*, 147(2):277–321, 2005.
143. J. Lee, J. Renz, and D. Wolter. Starvars - effective reasoning about relative directions. In F. Rossi, editor, *IJCAI*. IJCAI/AAAI, 2013.
144. S. Lemaignan, R. Ros, L. Mösenlechner, R. Alami, and M. Beetz. Oro, a knowledge management platform for cognitive architectures in robotics. In *IROS*, pages 3548–3553. IEEE, 2010.
145. H. Levesque. Planning with loops. In *Proc. IJCAI*, pages 509–515, 2005.
146. H. Levesque, R. Reiter, Y. Lespérance, F. Lin, and R. Scherl. Golog: A logic programming language for dynamic domains. *Journal of Logic Programming*, 31:59–84, 1997.
147. H. J. Levesque. Programming cognitive robots. https://www.cs.toronto.edu/ hector/pcr.html.
148. H. J. Levesque. All I know: a study in autoepistemic logic. *Artificial Intelligence*, 42(2-3):263–309, 1990.
149. H. J. Levesque. What is planning in the presence of sensing? In *Proc. AAAI / IAAI*, pages 1139–1146, 1996.
150. H. J. Levesque. *Common sense, the Turing test, and the quest for real AI*. Mit Press, 2017.
151. H. J. Levesque, E. Davis, and L. Morgenstern. The winograd schema challenge. In *Principles of Knowledge Representation and Reasoning: Proceedings of the Thirteenth International Conference, KR 2012, Rome, Italy, June 10-14, 2012*, 2012.
152. H. J. Levesque and G. Lakemeyer. *The logic of knowledge bases*. The MIT Press, 2001.
153. H. J. Levesque, F. Pirri, and R. Reiter. Foundations for the situation calculus. *Electron. Trans. Artif. Intell.*, 2:159–178, 1998.
154. D. Lewis. Probabilities of conditionals and conditional probabilities. *The philosophical review*, pages 297–315, 1976.
155. B. Limketkai, L. Liao, and D. Fox. Relational object maps for mobile robots. In *Proc. IJCAI*, pages 1471–1476, 2005.
156. F. Lin and H. J. Levesque. What robots can do: Robot programs and effective achievability. *Artif. Intell.*, 101(1-2):201–226, 1998.
157. F. Lin and R. Reiter. Forget it. In *Working Notes of AAAI Fall Symposium on Relevance*, pages 154–159, 1994.
158. F. Lin and R. Reiter. How to progress a database. *Artificial Intelligence*, 92(1-2):131–167, 1997.
159. D. Liu and G. Lakemeyer. Reasoning about beliefs and meta-beliefs by regression in an expressive probabilistic action logic. IJCAI, 2021.
160. Y. Liu and G. Lakemeyer. On first-order definability and computability of progression for local-effect actions and beyond. In *Proc. IJCAI*, pages 860–866, 2009.
161. Y. Liu and H. Levesque. Tractable reasoning with incomplete first-order knowledge in dynamic systems with context-dependent actions. In *Proc. IJCAI*, pages 522–527, 2005.
162. Y. Liu and X. Wen. On the progression of knowledge in the situation calculus. In *IJCAI*, 2011.
163. G. Marcus and E. Davis. *Rebooting AI: Building artificial intelligence we can trust*. Vintage, 2019.
164. Y. Martin and M. Thielscher. Integrating reasoning about actions and Bayesian networks. In *International Conference on Agents and Artificial Intelligence*, Valencia, Spain, Jan. 2009.

165. P. Mateus, A. Pacheco, J. Pinto, A. Sernadas, and C. Sernadas. Probabilistic situation calculus. *Annals of Math. and Artif. Intell.*, 32(1-4):393–431, 2001.

166. J. McCarthy. Programs with common sense. In *Proceedings of the Symposium on the Mechanization of Thought Processes*, National Physiology Lab, Teddington, England, 1958.

167. J. McCarthy and P. J. Hayes. Some philosophical problems from the standpoint of artificial intelligence. In *Machine Intelligence*, pages 463–502, 1969.

168. D. McDermott. Ray reiter's knowledge in action: a review. *AI Magazine*, 24(2):101–101, 2003.

169. B. Milch, B. Marthi, S. J. Russell, D. Sontag, D. L. Ong, and A. Kolobov. BLOG: Probabilistic models with unknown objects. In *Proc. IJCAI*, pages 1352–1359, 2005.

170. D. Monderer and D. Samet. Approximating common knowledge with common beliefs. *Games and Economic Behavior*, 1(2):170–190, June 1989.

171. R. C. Moore. A Formal Theory of Knowledge and Action. In *Formal Theories of the Commonsense World*, pages 319–358. Ablex, Norwood, NJ, 1985.

172. C. Muise, V. Belle, P. Felli, S. McIlraith, T. Miller, A. Pearce, and L. Sonenberg. Planning over multi-agent epistemic states: A classical planning approach. In *Proc. AAAI*, 2015.

173. K. Murphy. *Machine learning: a probabilistic perspective*. The MIT Press, 2012.

174. N. J. Nilsson. Probabilistic logic. *Artificial intelligence*, 28(1):71–87, 1986.

175. D. Nitti. *Hybrid Probabilistic Logic Programming*. PhD thesis, KU Leuven, 2016.

176. D. Nitti, V. Belle, and L. D. Raedt. Planning in discrete and continuous markov decision processes by probabilistic programming. In *ECML*, 2015.

177. D. Nitti, T. D. Laet, and L. D. Raedt. A particle filter for hybrid relational domains. In *IROS*, pages 2764–2771, 2013.

178. J. Pearl. *Probabilistic reasoning in intelligent systems: networks of plausible inference*. Morgan Kaufmann, 1988.

179. J. Pearl. *Causality*. Cambridge university press, 2009.

180. J. Pearl and D. Mackenzie. The book of why.

181. R. Petrick and F. Bacchus. Extending the knowledge-based approach to planning with incomplete information and sensing. In *Proc. ICAPS*, pages 2–11, 2004.

182. J. Pinto and R. Reiter. Reasoning about time in the situation calculus. *Annals of Mathematics and Artificial Intelligence*, 14(2):251–268, 1995.

183. R. Platt, L. P. Kaelbling, T. Lozano-Pérez, and R. Tedrake. Non-gaussian belief space planning: Correctness and complexity. In *ICRA*, pages 4711–4717, 2012.

184. D. Poole. First-order probabilistic inference. In *Proc. IJCAI*, pages 985–991, 2003.

185. I. Pratt-Hartmann. Total knowledge. In *Proc. AAAI*, pages 423–428, 2000.

186. M. L. Puterman. *Markov Decision Processes: Discrete Stochastic Dynamic Programming*. John Wiley & Sons, Inc., New York, NY, USA, 1st edition, 1994.

187. L. D. Raedt, K. Kersting, S. Natarajan, and D. Poole. Statistical relational artificial intelligence: Logic, probability, and computation. *Synthesis lectures on artificial intelligence and machine learning*, 10(2):1–189, 2016.

188. L. D. Raedt, A. Kimmig, and H. Toivonen. Problog: A probabilistic prolog and its application in link discovery. In *Proc. IJCAI*, pages 2462–2467, 2007.

189. R. Reiter. The frame problem in the situation calculus: a simple solution (sometimes) and a completeness result for goal regression. In *Artificial intelligence and mathematical theory of computation: Papers in honor of John McCarthy*, pages 359–380. Academic Press, 1991.

190. R. Reiter. *Knowledge in action: logical foundations for specifying and implementing dynamical systems*. MIT Press, 2001.

191. R. Reiter. On knowledge-based programming with sensing in the situation calculus. *ACM Trans. Comput. Log.*, 2(4):433–457, 2001.

192. M. Richardson and P. Domingos. Markov logic networks. *Machine learning*, 62(1):107–136, 2006.

193. S. J. Russell. Unifying logic and probability. *Commun. ACM*, 58(7):88–97, 2015.

194. T. Sang, P. Beame, and H. A. Kautz. Performing bayesian inference by weighted model counting. In *AAAI*, pages 475–482, 2005.

195. S. Sanner. Relational dynamic influence diagram language (rddl): Language description. Technical report, Australian National University, 2011.

196. S. Sanner and K. Kersting. Symbolic dynamic programming for first-order pomdps. In *Proc. AAAI*, pages 1140–1146, 2010.

197. M. Sap, V. Shwartz, A. Bosselut, Y. Choi, and D. Roth. Commonsense reasoning for natural language processing. In *Proceedings of the 58th Annual Meeting of the Association for Computational Linguistics: Tutorial Abstracts*, pages 27–33, 2020.

198. S. Sardina, G. De Giacomo, Y. Lespérance, and H. J. Levesque. On the semantics of deliberation in indigolog—from theory to implementation. *Annals of Mathematics and Artificial Intelligence*, 41(2-4):259–299, 2004.

199. S. Sardiña, G. De Giacomo, Y. Lespérance, and H. J. Levesque. On the limits of planning over belief states under strict uncertainty. In *KR*, pages 463–471, 2006.

200. S. Sardina and Y. Lespérance. Golog speaks the bdi language. In *Programming Multi-Agent Systems*, volume 5919 of *LNCS*, pages 82–99. Springer Berlin Heidelberg, 2010.

201. M. K. Sarker, L. Zhou, A. Eberhart, and P. Hitzler. Neuro-symbolic artificial intelligence. *AI Communications*, (Preprint):1–13, 2021.

202. R. B. Scherl and H. J. Levesque. Knowledge, action, and the frame problem. *Artificial Intelligence*, 144(1-2):1–39, 2003.

203. S. Shapiro, M. Pagnucco, Y. Lespérance, and H. J. Levesque. Iterated belief change in the situation calculus. *Artif. Intell.*, 175(1):165–192, 2011.

204. Y. Shoham. Agent-oriented programming. *Artificial intelligence*, 60(1):51–92, 1993.

205. P. Singla and P. M. Domingos. Markov logic in infinite domains. In *UAI*, pages 368–375, 2007.

206. R. Smullyan. *First-order logic*. Dover Publications, 1995.

207. W. Spohn. Ordinal conditional functions: A dynamic theory of epistemic states. In *Causation in Decision, Belief Change, and Statistics*, volume 42 of *The University of Western Ontario Series in Philosophy of Science*, pages 105–134. Springer Netherlands, 1988.

208. S. Srivastava. *Foundations and Applications of Generalized Planning*. PhD thesis, Department of Computer Science, University of Massachusetts Amherst, 2010.

209. L. Steels. The symbol grounding problem has been solved, so what's next? *Symbols and embodiment: Debates on meaning and cognition*, pages 223–244, 2008.

210. D. Suciu, D. Olteanu, C. Ré, and C. Koch. Probabilistic databases. *Synthesis Lectures on Data Management*, 3(2):1–180, 2011.

211. C. Swartz. *Introduction to gauge integrals*. World Scientific Publishing Company Incorporated, 2001.

212. S. Tellex, T. Kollar, S. Dickerson, M. R. Walter, A. G. Banerjee, S. Teller, and N. Roy. Approaching the symbol grounding problem with probabilistic graphical models. *AI magazine*, 32(4):64–76, 2011.

213. E. Ternovskaia. Inductive definability and the situation calculus. In B. Freitag, H. Decker, M. Kifer, and A. Voronkov, editors, *Transactions and Change in Logic Databases*, volume 1472 of *Lecture Notes in Computer Science*, pages 227–248. Springer Berlin Heidelberg, 1998.

214. M. Thielscher. From situation calculus to fluent calculus: state update axioms as a solution to the inferential frame problem. *Artificial Intelligence*, 111(1-2):277–299, 1999.

215. M. Thielscher. Planning with noisy actions (preliminary report). In *Proc. Australian Joint Conference on Artificial Intelligence*, pages 27–45, 2001.

216. M. Thielscher. Logic-based agents and the frame problem: A case for progression. In *First-Order Logic Revisited*, pages 323–336, Berlin, Germany, 2004. Logos.

217. M. Thielscher. Flux: A logic programming method for reasoning agents. *Theory and Practice of Logic Programming*, 5(4-5):533–565, 2005.

218. M. Thielscher. *Reasoning robots: the art and science of programming robotic agents*. Applied logic series. Springer, Dordrecht, 2005.

219. S. Thrun, W. Burgard, and D. Fox. *Probabilistic Robotics*. MIT Press, 2005.

220. V. B. Till Hofmann. Abstracting noisy robot programs. In preparation, 2022.

221. S. D. Tran and L. S. Davis. Event modeling and recognition using markov logic networks. In *Proc. ECCV*, pages 610–623, 2008.

222. W. F. Trench. *Introduction to real analysis*. Prentice Hall, 2003.

223. L. Treszkai and V. Belle. A correctness result for synthesizing plans with loops in stochastic domains. *International Journal of Approximate Reasoning*, 119:92–107, 2020.

224. J. Van Benthem, J. Gerbrandy, and B. Kooi. Dynamic update with probabilities. *Studia Logica*, 93(1):67–96, 2009.

225. G. Van den Broeck. *Lifted Inference and Learning in Statistical Relational Models*. PhD thesis, KU Leuven, 2013.

226. H. Van Ditmarsch, A. Herzig, and T. De Lima. Optimal regression for reasoning about knowledge and actions. In *Proc. AAAI*, pages 1070–1075, 2007.

227. H. van Ditmarsch, W. van der Hoek, and B. Kooi. *Dynamic Epistemic Logic*. Springer Publishing Company, Incorporated, 1st edition, 2007.

228. H. P. van Ditmarsch, A. Herzig, and T. D. Lima. From situation calculus to dynamic epistemic logic. *J. Log. Comput.*, 21(2):179–204, 2011.

229. S. Vassos and H. Levesque. On the Progression of Situation Calculus Basic Action Theories: Resolving a 10-year-old Conjecture. In *Proc. AAAI*, pages 1004–1009, 2008.

230. J. Vennekens, M. Bruynooghe, and M. Denecker. Embracing events in causal modelling: Interventions and counterfactuals in cp-logic. In *European Workshop on Logics in Artificial Intelligence*, pages 313–325. Springer, 2010.

231. R. Waldinger. Achieving several goals simultaneously. In *Machine Intelligence*, volume 8, pages 94–136. 1977.

232. H. Younes and M. Littman. PPDDL 1. 0: An extension to pddl for expressing planning domains with probabilistic effects. Technical report, Carnegie Mellon University, 2004.

233. Z. Zamani, S. Sanner, P. Poupart, and K. Kersting. Symbolic dynamic programming for continuous state and observation POMDPs. In *NIPS*, pages 1403–1411, 2012.

234. B. Zarrieß and J. Claßen. Verification of knowledge-based programs over description logic actions. In Q. Yang and M. Wooldridge, editors, *Proceedings of the Twenty-Fourth International Joint Conference on Artificial Intelligence (IJCAI 2015)*, pages 3278–3284. AAAI Press, 2015.

Printed in the United States
by Baker & Taylor Publisher Services